F/A-18 HORNET

Walter J. Boyne Military Aircraft Series

F-22 Raptor
America's Next Lethal War Machine
STEVE PACE

B-1 Lancer
The Most Complicated Warplane Ever Developed
DENNIS R. JENKINS

B-24 Liberator
Rugged but Right
FREDERICK A. JOHNSEN

B-2 Spirit
The Most Capable War Machine on the Planet
STEVE PACE

F/A-18 Hornet
A Navy Success Story
DENNIS R. JENKINS

B-17 Flying Fortress
The Symbol of Second World War Air Power
FREDERICK A. JOHNSEN

F-105 Thunderchief
Workhorse of the Vietnam War
DENNIS R. JENKINS

B-47 Stratojet
Boeing's Brilliant Bomber
JAN TEGLER

F/A-18 HORNET

A Navy Success Story

Dennis R. Jenkins

McGraw-Hill

New York   San Francisco   Washington, D.C.   Auckland   Bogotá
Caracas   Lisbon   London   Madrid   Mexico City   Milan
Montreal   New Delhi   San Juan   Singapore
Sydney   Tokyo   Toronto

Library of Congress Cataloging-in-Publication Data

Jenkins, Dennis R.
    F/A-18 Hornet : a Navy success story / Dennis R. Jenkins.
        p.   cm.
    Includes bibliographical references and index.
    ISBN 0-07-134696-1
    1. Hornet (Jet fighter plane) I. Title.
  UG1242.F5 J46   2000
  623.7'464—dc21   99-087128
                                               CIP

McGraw-Hill

A Division of The McGraw-Hill Companies

  2 3 4 5 6 7 8 9 0 KGP/KGP 0 6 5 4 3 2 1

ISBN 0-07-134696-1

*The sponsoring editor for this book was Shelley Ingram Carr, the editing
supervisor was Stephen M. Smith, and the production supervisor was Pamela
A. Pelton. It was set in Utopia by North Market Street Graphics.*

Printed and bound by Quebecor/Kingsport.

CONTENTS

The McGraw-Hill Companies is pleased to present the **Walter J. Boyne Military Aircraft Series.** The series will feature comprehensive coverage, in words and photos, of the most important military aircraft of our time.

Profiles of aircraft critical to defense superiority in World War II, Korea, Vietnam, the Cold War, the Gulf War, and future theaters detail the technology, engineering, design, missions, and people that give these aircraft their edge. Their origins, the competitions between manufacturers, the glitches and failures, and type modifications are presented along with performance data, specifications, and inside stories.

To ensure that quality standards set for this series are met volume after volume, McGraw-Hill is immensely pleased to have Walter J. Boyne on board. In addition to his overall supervision of the series, Walter is contributing a Foreword to each volume that provides the scope and dimension of the featured aircraft.

Walter was selected as editor because of his international preeminence in the field of military aviation and particularly in aviation history. His consuming, lifelong interest in aerospace subjects is combined with an amazing memory for facts and a passion for research. His knowledge is enhanced by his personal acquaintance with many of the great pilots, designers, and business managers of the aviation industry.

As a Command Pilot in the United States Air Force, Colonel Boyne flew more than 5000 hours in a score of different military and civil aircraft. After his retirement from the Air Force in 1974, he joined the Smithsonian Institution's National Air & Space Museum, where he became Acting Director in 1981 and Director in 1986. Among his accomplishments at the Museum were the conversion of Silver Hill from total disarray to the popular and well-maintained Paul Garber Facility, and the founding of the very successful *Air&Space/ Smithsonian* magazine. He was also responsible for the creation of NASM's large, glass-enclosed restaurant facility. After obtaining permission to install IMAX cameras on the Space Shuttle, he supervised the production of two IMAX films. In 1985, he began the formal process that will lead ultimately to the creation of a NASM restoration facility at Dulles Airport in Virginia.

Boyne's professional writing career began in 1962; since that time he has written more than 500 articles and 28 books, primarily on aviation subjects. He is one of the few authors to have had both fiction and nonfiction books on *The New York Times* best seller lists. His books include four novels, two books on the Gulf War, one book on art, and one on automobiles. His books have been published in Canada, Czechoslovakia, England, Germany, Italy, Japan, and Poland. Several have been made into documentary videos, with Boyne acting as host and narrator.

Boyne has acted as consultant to dozens of museums around the world. His clients also include aerospace firms, publishing houses, and television companies. Widely recognized as an expert on aviation and military subjects, he is frequently interviewed on major broadcast and cable networks, and is often asked by publishers to review manuscripts and recommend for or against publication.

Colonel Boyne will bring his expertise to bear on this series of books by selecting authors and titles, and working closely with the authors during the writing process. He will review completed manuscripts for content, context, and accuracy. His desire is to present well-written, accurate books that will come to be regarded as definitive in their field.

Author Dennis R. Jenkins has rendered what I consider to be the best aircraft monograph ever written in his definitive *F/A-18 Hornet: A Navy Success Story.* He has given us an encyclopedic account of an immensely sophisticated weapon system, the Hornet, in all its many variations. More than this, he has put the Hornet, its rivals and competitors, its weapons, and its subsystems into context, so that his story also contains a succinct history of fighter aircraft development for the last 40 years.

Jenkins backs up the text with a marvelous collection of photographs, each one with a pertinent caption that further illuminates the work. Taken all together, it is a stunning achievement.

The Hornet was both an ideal and a terrible subject for a book, and it might have posed problems for someone less versed in the subject. It was ideal in that it represented a new generation of Navy fighters, the product of the ferment of fighter development in the 1960s and 1970s. It was terrible because the many complexities of the program could have been extraordinarily difficult to follow without Jenkins's masterful handling.

Many authors, myself included, have a tendency to fall in love with the airplane they are describing, and faithfully record every positive feature while glossing over anything negative. Jenkins avoids this with dispassionate exactitude, recording not only the many successful innovations introduced in the Hornet, but also noting where trouble occurred and how it was surmounted. It is in these deft accounts that Jenkins excels; his writing draws a sharp, clearly delineated picture of the problem, and then, with equal skill, depicts how the problem was solved. For example, he writes of problems encountered during early testing, when the Hornet's takeoff run was excessively long, as follows:

> These problems were solved by filling in the snag on the leading edge of the stabilators, giving them greater authority at an earlier juncture during the takeoff run. The snag had been added to the leading edge of the stabilator in anticipation of the same flutter problems that had affected the F-15, but these problems never materialized. In addition, a greater upward moment was provided by automatically toeing in the rudders during takeoff, reducing rotation speed, by 30 knots, to 115 knots.

In that one paragraph, he describes a problem, its history, the fix, and the results, and this easily read but comprehensive analysis is typical of the entire book.

Jenkins sets the tone for this book with an insightful portrayal of the history of that often sought but rarely found aircraft, the lightweight fighter. His account of the so-called Fighter Mafia within the Air Force and its effect upon the development of the F-15, F-16, and F-18 fighters is the best—and shortest—that I've ever read, showing, as it does, how a band of dedicated activists can prevail even in the Pentagon.

In a similar way, the author details just how the F/A-18 was born of the YF-16/YF-17 competition, and gives an explicit account of the later corporate problems between Northrop and McDonnell Douglas when it came to the exploitation of the basic design. Jenkins's ability to give a comprehensive account in a minimum number of words is the key to his successful coverage of the Hornet, which includes aircraft-by-aircraft assignment to fleet squadrons and to specialized domestic use.

He deals with the Hornet's longstanding insufficient-range problems in the same way, laying out in exact terms what the deficiencies are and what corrective measures have been taken. It is in matters like these that his broad knowledge of the subject matter comes into play, for Jenkins comprehends the changing nature of the Navy's role, from confrontation with the Soviet Union to policing littoral waters of Third World states. He deals with the question of stealth, and the measure of the Hornet's adaptation to the requirement, with consummate skill.

This book provides broad insight into the use of the Hornet by foreign countries (and also those instances where the F/A-18 lost out to competitors) and also gives comprehensive coverage of its use in combat.

For many readers, the most illuminating part of the book will be Jenkins's adroit description of the similarities of the F/A-18E/F Super Hornet with earlier versions of the aircraft and the differences between them. The Super Hornet is, of course, an almost totally new airplane in terms of its structure and size, but the similarity in appearance and in nomenclature with previous incarnations has obscured this. In addition, incremental improvements to the F/A C/D aircraft have further muddied the waters. Jenkins explains the situation exactly and, in addition, offers some thoughtful comments on the aircraft's future production.

Finally, for the true devotee, the author provides unusual insight into the complex electronic systems as well as the many ordnance systems that the Hornet series can use. I think these explanations are at once the simplest and most revealing of any that I have read, and just as the rest of the book does, provide a standard for other authors to match. *F/A-18 Hornet: A Navy Success Story* is also a Dennis Jenkins success story.

—Walter J. Boyne

The F/A-18 Hornet should be suffering from an identity crisis. What began as a simple Air Force lightweight fighter prototype has evolved into the most sophisticated warplane in the U.S. Navy inventory. Along the way, there have been three distinct aircraft that have shared the basic Hornet configuration. First were the two Northrop YF-17 Cobra demonstrators. Next came the production Hornets—over 1,400 of them for eight different customers. The third variant, just entering service, is the Super Hornet.

In many respects the Hornet is unique. It evolved largely from an Air Force design, one of the few types to successfully make the transition. It is one of the few carrier-based aircraft to enjoy a large export market—only the Douglas A-4 Skyhawk has been more widely exported. By all accounts the Hornet is a capable aircraft, one thoroughly enjoyed by its pilots and maintenance crews. But it has been largely overshadowed by the public's enthrallment with the F-14 and F-16 thanks to movies such as "Top Gun" and "Iron Eagle."

The 1990s brought many changes. One of the more frustrating for historians was the megamergers that occurred within the defense industry in the United States. It started with "teaming"—multiple companies agreeing to put competition aside and co-produce an aircraft. The original YF-17 was developed by the Northrop Corporation, but an agreement was reached that McDonnell Douglas would be prime contractor for any naval versions, that company having more of a reputation within the Navy. In the end, only the naval version was ordered into production, and Northrop was relegated to a role as a major subcontractor on an aircraft known as the McDonnell Douglas F/A-18 Hornet. Then McDonnell Douglas merged with the Boeing Company on 4 August 1997. So the subject of this monograph is now the Boeing F/A-18 Hornet. The original Hornet industry team consisted of McDonnell Douglas, Northrop, Hughes, and General Electric. Interestingly, in 2000 only GE retains its traditional name, the rest of the team becoming Boeing, Northrop Grumman, and Raytheon, respectively.

For those readers wishing a more detailed history of the Northrop YF-17 prototypes, Don Logan's excellent *Northrop's YF-17 Cobra* (Schiffer Publishing, Ltd., Atglen, Pa., 1996) is probably the ultimate work on the subject.

Thanks go to my friends Mick Roth, Tony Landis, Frederick A. Johnsen, Terry Panopalis, and Kev Darling for their assistance with this project. Ellen LeMond-Holman at Boeing provided many excellent photos and coordinated a review of the manuscript within Boeing (thanks to Larry Merritt, Lon O. Nordeen, Jr., Frank Vertovek, Gil Rud, Tim Staley, and Bob Goodwin). Ned Conger also coordinated a review with Admiral John Weaver (Ret.), former F/A-18 program manager and now chairman and CEO of Raytheon International; Captain (Rear Admiral–select) James B. Godwin III, Navy F/A-18 program manager; Captain Jeffrey

Wieringa, Navy F/A-18 deputy program manager; Frank Amorosi; and John T. "Tom" Lindgren. Special thanks go to former F/A-18 test pilots Ivan Behel and Jim Sandberg who provided some extremely insightful comments. After the extensive reviews, any mistakes that remain in this book are purely my fault.

Mike Reyno at Skytech Images provided some absolutely outstanding photography of CF-18s, including the cover shot. Chris B. Hake from No. 3 Squadron provided some insight into the RAAF Hornets, and obtained some excellent photos from Leading Aircraftsmen Jason Freeman and Peter Gammie. Leena Tiainen, the Information Secretary at the Finnish Air Force Headquarters, graciously contributed several excellent photographs. Rita Berchtold from the Swiss Air Force information office went out of her way in assisting Neil Dunridge during his visit to photograph the Swiss Hornets for this book. Felicia A. Campbell at Raytheon Systems graciously provided photos. Many others also contributed, including Don Logan, Cristoph Kugler, Mark Munzel, Mike Grove, Bill Kistler, J. T. Wenting, Patrick Nieuwkamp, Peter Merlin, and Jeff Thurman at the Defense Visual Information Services. The whole project would not have been possible without Shelley Carr at McGraw-Hill and Walt Boyne, who lent his name and expertise to the series. On the personal side, many thanks are given to Cyndy Thomas and my mother, Mary E. Jenkins, for putting up with me while I spent nights figuring out how to write this book.

—DENNIS R. JENKINS

An Unusual Beginning: The Lightweight Fighter Competition

Surprisingly for a naval aircraft, the F/A-18 can trace its origins to an Air Force fighter design.[1] During the 1960s the Air Force was seeking a new fighter, and the resulting F-X competition was eventually won by the McDonnell Douglas F-15 Eagle. The F-15 has largely validated the usefulness of a large heavyweight fighter, racking up an impressive tally of air-to-air victories in the service of the United States, Israel, and Saudi Arabia. During the competition, however, there was a small, but vocal, group of Air Force officers and Department of Defense civilians that argued against large fighters.[2]

During the Korean War, the United States Air Force had racked up a 7:1 kill-to-loss ratio using North American F-86 Sabres, mainly against Soviet-built MiG-15s. The F-86 had first flown in May 1947, and looked every bit a fighter in the tradition of the P-51 Mustang. In fact, the Sabre was the only U.S. jet aircraft to see significant service in Korea that had been designed primarily for aerial combat, albeit as a fighter-escort. All of the others had been designed either as fighter-bombers or high-speed interceptors.

Project Forecast, an attempt by the Air Force during 1963 to identify future weapons requirements, foresaw such notable developments as the C-5 and B-1 programs. Directed by General Bernard A. Schriever, commander of the Air Force Systems Command, Project Forecast proved somewhat less clairvoyant regarding future fighter programs. It predicted that Air Force fighter needs for the next 20 years would be met by missile-armed variants of the F-111 and F-4 ". . . optimized for the air-superiority role," and that strategic bombing would be conducted from aircraft able to fly faster and higher than the enemy. Almost as an afterthought, Project Forecast added that a ". . . counterair force must be able to destroy aircraft in the air . . ." at long ranges using advanced weapon systems. And so, American designers got sidetracked from developing true fighters by the magic of radar and electronics. These, coupled with the advent of seemingly workable beyond-visual-range (BVR) missiles, made the traditional "dogfight" appear to be obsolete. All future battles would be fought without ever seeing the enemy, or so it was thought.

The first opportunity to use this new technology, and an entire generation of aircraft built around it, came in Vietnam. It did not work. Not only were the new electronics and

The Northrop N-102 FANG was a lightweight day fighter that could be easily converted into a fighter-bomber. The highly-swept delta wing suggests the Mach 2 capabilities of the design. Note the belly air intake, similar to that on the eventual F-16. *(Northrop)*

missiles unreliable, the entire battle scenario was vastly different from the exercises of the late 1950s and early 1960s. The major problem was that the enemy did not cooperate. During the "war games," the bad guys had always approached from one side, the good guys from the other. All the blips on the radar "over there" were the enemy, and fair game to fire at. The real world didn't work that way, since friends and enemies were interspersed in the same air space. To tell the players apart, identification, friend or foe (IFF) systems were improved upon, but they were still unreliable, forcing the pilots to make visual identification prior to engaging. These problems with tactics and equipment, coupled with unbelievable political considerations, conspired against the American fighter pilot and created the rather dismal 2.5:1 Air Force kill ratio in Vietnam.

The *Force Options for Tactical Air* study had been initiated in August 1964 under Lieutenant Colonel John W. Bohn, Jr., to assess the Air Force's reliance on high-cost, high-performance tactical fighters. Completed on 27 February 1965, the study concluded that aircraft such as the F-111 were far too costly to be risked in a limited (i.e., nonnuclear) war and recommended the acquisition of a mix of high- and low-cost aircraft as the most economical method of strengthening the tactical force. For the low-cost role, the study narrowed the candidates to the comparatively inexpensive Northrop F-5 Freedom Fighter and the Navy's Ling-Temco-Vought (LTV) A-7 Corsair II. Both seemed equally acceptable—the A-7 could carry a greater payload and offered commonality with the Navy, whereas the F-5 had superior air-to-air combat capabilities. For a variety of reasons, the high-low mix suggestion was not acted upon.[3]

The Air Force began studying the basis for a new fighter during April 1965. The Fighter-Experimental (F-X) would possess ". . . superior air-to-air, all-weather . . ." capabilities, and was envisioned as a single-seat, twin-engine fighter stressing maneuverability over speed, with an initial operational capability (IOC) in 1970. As a result of this study, on 6 October 1965, the Tactical Air Command (TAC) released Qualitative Operations Requirement (QOR) 65-14-F outlining the need for a new air-superiority fighter emphasizing an ". . . aircraft capable of out-performing the enemy in the air. . . ." Other desired features included a high thrust-to-weight ratio, an advanced air-to-air radar, a top speed of Mach 2.5, and armament consisting of infrared short-range and radar-guided BVR missiles.

The Northrop family of lightweight aircraft on display in Hawthorne on 4 April 1974. The YF-17 is in the foreground, an F-5E is in camouflage, and an F-5B is slightly behind it. The back row is an F-5A and T-38 (white). The YF-17 design was a fairly large departure from the extremely successful F-5/T-38 family. Note the large boundary layer air discharge (BLAD) on the leading-edge extension (LEX) of the YF-17, a feature that would disappear from the F/A-18. *(Northrop via Don Logan)*

During the summer and fall of 1965, the Air Force came back to Bohn's F-5 versus A-7 issue. The Office of the Secretary of Defense (OSD) was still enamored with the concept of "commonality," where the Air Force and the Navy would have a force comprising F-111, F-4, and A-7 aircraft. In July, Secretary of Defense Robert S. McNamara directed the Air Force to select the A-7 for the close air support role, or demonstrate a compelling reason not to. At the same time, but on a lower priority, he endorsed the Air Force's work on the F-X.

On 5 November the new Secretary of Defense, Dr. Harold Brown, and Air Force Chief of Staff General John P. McConnell proposed acquiring 11 squadrons (264 aircraft) of A-7s. Although criticized in some Air Force circles as a capitulation, the decision to buy the A-7 was in fact a sensible compromise that ultimately cleared the way for approval of the F-X, which could now be justified as a "... more sophisticated, higher-performance, air-craft ... as an air-superiority replacement for the F-4...."

On 8 December 1965 a request for proposals (RFP) was released to 13 manufacturers for initial parametric design studies of the F-X. The RFP specified an aircraft that combined air-to-air and air-to-ground capabilities, despite the widely held belief that what was really needed was a dedicated air-superiority fighter. Proposals were received from eight companies, and on 18 March 1966 contracts were awarded to Boeing, Lockheed, and North American for a 4-month concept formulation study.[4] One other company, Grumman, participated in the study on an unfunded basis.

After considering the effects of five variables—avionics, maneuverability, payload, combat radius, and speed—on the F-X in terms of weight and cost, the contractors came up with some 500 possible designs. The typical concept weighed more than 60,000 pounds, had variable-geometry wings, and required the use of exotic materials to obtain a top speed of Mach 2.7. The aircraft more closely resembled the F-111 than the F-86.

Energy Maneuverability

The Air Force was not satisfied with the results of the study, and did not further pursue the proposals. Since the final RFP had specified an aircraft capable of both air-to-air and air-to-ground operations, what emerged was one optimized for neither. The Air Force Systems Command sensed that the F-X requirements were ". . . badly spelled out . . ." and subsequently persuaded TAC to modify the requirements. This was largely due to the work of Major John R. Boyd.[5]

In October 1966 Boyd had joined the Tactical Division of the Air Staff Directorate of Requirements, and, when asked to comment on the just-completed F-X proposals, he summarily rejected the designs as inappropriate to the task. A veteran combat pilot of the late 1950s and author of the air combat training manual used by the Fighter Weapons School at Nellis Air Force Base (AFB), Boyd was well qualified to assess fighter aircraft. In 1962, as part of an engineering course at Georgia Tech, he had studied the energy changes incurred by an aircraft during flight. Boyd hypothesized that a fighter's performance at any combination of altitude and airspeed could be expressed as the sum of its potential and kinetic energies, and its capability to change these energy states by maneuvering. With this idea as a departure point, Boyd believed he could describe how well a fighter could perform at any point in its flight envelope. If the hypothesis were true, this could be used to compare the performance of competing aircraft and determine which one was superior at any point in the flight envelope.

By establishing a standard for comparison, Boyd accomplished two significant things: He could compare the performance characteristics of existing fighters and derive better combat tactics, and he could evolve an improved design for a truly superior fighter. Although elegant in its simplicity, and computationally straightforward, Boyd's energy-maneuverability theory required millions of calculations, something that could be accomplished only by the large mainframe computers of the 1960s. But computer time was expensive and Boyd had no official authorization or budget to pursue his theory.

Nevertheless, Boyd continued his energy-maneuverability studies at his next assignment even though his billet was as a maintenance officer. At Eglin AFB he met Thomas Christie, a mathematician who also saw promise in the energy-maneuverability theory and who had access to a high-speed computer. With Christie's help, Boyd concocted an elaborate cover story to disguise the actual use of the computer time. Initial comparisons centered on the MiG-15 and F-86 used during the Korean War. Later work included comparing MiGs with the F-4 used in Vietnam.

Much to the dismay of many, including the aerodynamic engineers at Wright Field, Boyd's energy-maneuverability theory proved to be quite accurate. It provided a convenient method for translating tactics into engineering specifications and vice versa. Although the energy-maneuverability theory did not represent anything new in terms of physics or aerodynamics, it provided planners and developers a formalized method to compare competing aircraft directly, and to demonstrate the effects of design changes on aircraft performance. The energy-maneuverability theory would revolutionize the way designers looked at fighters. On the basis of this method, a model was designed that would demonstrate the effects of specific requirements on the F-X design. By the spring of 1967, through the efforts of Boyd and others, the projected weight of the F-X had been reduced

The McDonnell Douglas F-15 Eagle was the Air Force's fighter of choice, but budgetary considerations eventually led to the large-scale production of a smaller, lighter, and less expensive aircraft—the F-16. This F-15 has a full load of air-to-air missiles: four AIM-7 Sparrows on the fuselage stations and four AIM-9 Sidewinders under the wings. Like the Hornet, the F-15 proved to have slightly short legs and is usually seen with a centerline fuel tank. *(U.S. Air Force)*

from 60,000+ pounds to slightly under 40,000 pounds, and the top speed scaled down to Mach 2.3 to 2.5.

Surprises at Domodadovo

The F-X formulation phase continued through the middle of 1967. Then, in July 1967, the Russians held the famous Domodadovo air show where they introduced six new aircraft types, including the MiG-25 Foxbat,[6] as well as several new versions of older aircraft. Soon afterward, the Air Force submitted the F-X proposal to OSD as the new tactical fighter to replace the F-4. The Air Force argued for the importance of air superiority, without which other aerial missions (close air support and interdiction)[7] would be either too costly or impossible to conduct. It was noted that although the multipurpose F-4 was a capable air-to-air fighter, its continued effectiveness was doubtful in view of the new Soviet fighters.[8]

On 11 August 1967 the Air Force solicited bids from seven contractors for a second round of F-X concept studies. Contracts were awarded to General Dynamics and McDonnell Douglas on 1 December 1967 with four other companies, Fairchild-Republic, Grumman, Lockheed, and North American, participating with company funds. The study concluded in June 1968, and was reviewed by an Air Force team that numbered more than 100. The basic airframe issues were resolved quickly, but the composition of the avionics suite caused considerable disagreement. Specifically, the multipurpose advocates attempted to retain such items as terrain-following radar and blind-bombing capabilities. They argued that future advances in technology would permit weight reductions acceptable to the F-X, but overlooked the costs and risks involved.

Like the Air Force's F-15, the Navy's Grumman F-14 Tomcat is a very large, capable fighter. But it too was expensive, and it also had the bad fortune of encountering some development problems that almost caused Grumman to go out of business. The Navy soon realized that it too would need a lightweight fighter to complement the Tomcat. *(Robert L. Lawson via the Jay Miller Collection)*

Several significant events occurred in 1968 that helped shape the course of the F-X program. The Navy had become disenchanted with the joint Tactical Fighter, Experimental (TFX) program, and had initiated the development of the Naval Fighter-Experimental (VFX)—which became the Grumman F-14 Tomcat. Also, the presidential election in November guaranteed a change in the civilian leadership, both in the White House and the Pentagon. Since the Department of Defense still favored the concept of "common" hardware, the Air Force decided to make the requirements for the F-X sufficiently different from the Navy's VFX to justify continued development. And, in an effort to get far enough along in development to protect the program from cancellation by the new administration, the Air Force decided to skip the prototype phase and proceed directly to full-scale development, a ploy also used by the F-14 program. In May 1968 the F-X became the Air Force's top priority development program.

The Air Force described the F-X as a ". . . single-seat, twin-engine aircraft featuring excellent pilot visibility, with internal fuel sized for a 260-nautical-mile design mission, and . . . a balanced combination of standoff [missile] and close-in [gun] target kill potential." The decision to include just one crew member was arrived at as much to differentiate the aircraft from the Navy's VFX as to save the estimated 5,000 pounds in additional structure and systems. The twin engine design was selected because it featured faster throttle response and earlier availability (interestingly, safety does not seem to have been a factor).

In a letter dated 12 September 1968, the Air Force requested a designation for the new fighter. The Navy had earlier rejected the next available fighter designation (F-13) in favor of F-14 for the VFX. With superstition apparently influencing the decision again, the Air Force also declined the F-13 designation and requested F-15 instead.

Dissension within the Air Force

Not all in the Air Force agreed that the F-15 was the aircraft to buy. One proposed alternative, dubbed the F-XX, was offered by Pierre M. Sprey of the OSD Office of Systems Analysis during July 1968. He believed that the F-15 was too expensive, incorporated too much high-risk technology, was unnecessarily complex, and would not achieve its advertised air superiority performance. This point of view was shared by Boyd. Along with Chuck Meyers, Everest Riccioni, and others, Sprey and Boyd became known as the "Fighter Mafia." Sprey's alternative was a 25,000-pound, single-seat, single-engine aircraft designed specifically to fight in the transonic region. Sprey's F-XX proposal shunned complex avionics, featuring instead a simple gunsight radar and straightforward maintenance. The proposal also included a VF-XX version for the Navy.[9]

The Air Force and Navy were not interested. They cited the short, unhappy experience of similarly equipped F-104s in Vietnam and the limitations of the F-5 as examples of the inadequacy of lightweight fighters. But the Fighter Mafia was not alone in advocating lightweight fighters. Indeed, many experienced Air Force and Navy fighter pilots recommended that the best solution to the air-superiority problem was to ". . . buy MiG-21s!" Simulations and flight tests during 1968 (including Projects Feather Duster and Have Doughnut) demonstrated the superior maneuverability of a lightweight fighter against F-4Es. Although the idea had considerable merit, it was ill-timed. The F-14 and F-15 development programs

The F-14 was introduced to the fleet when colorful markings were still allowed, and VF-1 Wolfpack made full use of the opportunity with these striking red markings. Almost 30 years after the F-14 first entered service it is still one of the world's most capable long-range air defense fighters, thanks mainly to its APG-71 radar and AIM-54 Phoenix missiles. Recently it has begun to acquire a useful air-to-ground capability as well. *(Robert L. Lawson via the Jay Miller Collection)*

were too far along to be sidetracked, and all the proposal succeeded in doing was uniting the services behind their respective heavyweight fighters.

Lightweight Fighters

But the seeds of a new concept had been planted, and they quickly took root once the F-15's development was assured. As early as 1965 the Air Force had begun low-priority studies of a lightweight Advanced Day Fighter (ADF). The ADF was to be in the 25,000-pound class and was to have a thrust-to-weight ratio and wing loading intended to better the performance of the MiG-21 by at least 25 percent. However, the ADF was temporarily shelved after the 1967 Domodadovo air show so that all resources could be concentrated on the F-15.[10]

But the ADF concept was kept alive by the Fighter Mafia, and a 1969 Pentagon memorandum suggested that both services adopt the F-XX as a substitute for the F-15 and F-14 since both aircraft were becoming increasingly expensive. The services vigorously resisted, and both the F-14 and F-15 programs continued. Nevertheless, both services realized that the increasing costs of their desired fighters, combined with decreasing real-world defense budgets, would eventually force them to consider alternatives.

Deputy Secretary of Defense David A. Packard (who came in with the new Nixon Administration in 1969) was a strong advocate of returning to the concept of competitive prototyping as a way of containing the ever-increasing costs of new weapons systems. By 1971, Boyd was working for the Air Force Prototype Study Group and was able to push the ADF concept at a time when the idea of competitive fly-offs was coming back into fashion. Packard was swayed by the Fighter Mafia's arguments, and endorsed Boyd's concept. Some engineers within the aerospace industry, specifically Lockheed, Northrop, and Vought, also supported the concept of a small inexpensive fighter.[11]

The result of this lobbying was that Congress set aside $12 million in the FY72 budget specifically for the Light Weight Fighter (LWF) program. An RFP was issued to industry on 6 January 1972, calling for a small day fighter with a high thrust-to-weight ratio, a gross weight of less than 20,000 pounds, and high maneuverability. Although generally similar to the ideas presented by Boyd and Sprey, the requested aircraft was lighter than either the ADF or F-XX concepts had proposed.

The aircraft was intended to complement the F-15, not replace it. No attempt would be made to equal the performance of the MiG-25, the aircraft being designed instead for the most likely conditions of future air combat—altitudes of 30,000 to 40,000 feet and speeds of Mach 0.6 to 1.6. Emphasis was to be on maneuverability and acceleration rather than on high speed or weapons-carrying ability. A small size was stressed, since this had made the MiG-17 and MiG-21 difficult to detect visually during combat over North Vietnam. The RFP specified three main objectives—the aircraft should fully explore the advantages of emerging technologies, reduce the risk and uncertainties involved in full-scale development and production, and provide a variety of technological options to meet future military hardware needs.[12]

Five manufacturers—Boeing, General Dynamics, Lockheed, Northrop, and Vought—submitted proposals by the 18 February 1972 deadline. In March 1972, the Air Staff concluded that the Boeing Model 908-909 was most desirable, with the General Dynamics Model 401-16B and Northrop Model P600 being rated as close seconds. The Vought V-1100 and Lockheed CL-1200 Lancer had been eliminated.

Further evaluation rated the General Dynamics and Northrop proposals ahead of the Boeing submission. This was based partially on the fact that the Boeing and General Dynamics concepts were quite similar in overall appearance and technology, and the General Dynamics proposal was less expensive. Since one of the goals of the program was to validate emerging technologies, the decision was to select two substantially different approaches. The final decision was made by Secretary of the Air Force Robert Seamans.[13]

On 13 April 1972, contracts for two General Dynamics YF-16s (72-1567/1568) and two Northrop YF-17s (72-1569/1570) were awarded. General Dynamics would receive $37.9 million, while Northrop would be paid $39.9 million. The "cost plus fixed fee" contracts covered the design and construction of two prototypes, plus 300 hours of flight testing. At the same time, contracts were awarded to Pratt & Whitney for a version of the F-15's F100 turbofan specially adapted for single-engined aircraft, and to General Electric for the YJ101 engine.[14]

Encouraged by the contract awards, the Fighter Mafia continued to push for a production commitment. The Air Force was more concerned at the time about bringing the F-15 into full production, a fact made somewhat more difficult by early teething problems with the F100 engine. But the inflationary spiral that characterized the U.S. economy in the early 1970s was beginning to catch up with large defense programs. It was being quietly admitted that the Air Force would not be able to afford as many F-15s as it wanted, and the "high-low" mix first discussed in 1964 was soon embraced.

Under this idea the Air Force would purchase as many of the expensive, highly capable, F-15s as Congress would allow. A smaller, less expensive but much less capable, fighter would also be purchased to bring the Air Force up to the desired strength. It was reasoned that once the F-15s had achieved basic air superiority over a given enemy, the smaller fighter could be used with relative impunity. The end result was that the LWF competition, which had started as a technology/concept demonstration program, was beginning to look like a competition for a large-scale production contract.

The General Dynamics Entry

The YF-16 was designed and built at Fort Worth under the direction of William C. Dietz and Lyman C. Josephs, with Harry Hillaker as chief designer. Proved systems and components were used throughout most of the aircraft, and components and detail assemblies were

The two YF-16 prototypes early in the flight demonstration program. The first prototype (72-1567) broke with tradition and was painted in red, white, and blue markings, making it look even sleeker than it already did. The second prototype (72-1568) was initially painted in a blue and white "broken sky" camouflage, but was quickly repainted an overall gray. *(Lockheed Martin Tactical Air Systems via Mike Moore)*

designed for ease of manufacture with low-cost materials where possible. However, the F-16 incorporated an analog fly-by-wire flight control system and relaxed static longitudinal stability, both of which represented significant advances in fighter technology. Equally unique, the pilot sat in a highly reclined position and was provided with a side-stick controller instead of a conventional control stick.

General Dynamics elected to use a single Pratt & Whitney F100 turbofan, since it was estimated to have substantially lower fuel consumption than a pair of YJ101s, and studies revealed no significant attrition advantage for a twin-engine arrangement.[15] The single-engine design made it possible to achieve a mission weight of 17,050 pounds, whereas General Dynamics estimated an aircraft powered by two YJ101s would have had a mission weight of 21,470 pounds. It was not far off.

The first YF-16 (72-1567) was rolled out at Fort Worth on 13 December 1973 and was delivered by C-5A to Edwards AFB on 8 January 1974. The aircraft's first flight was an unintended short hop around the pattern on 21 January 1974 at the hands of Phil Oestricher. During high-speed ground tests violent lateral oscillations resulted in the scraping of the right stabilator on the runway. Wisely, Oestricher decided to take off, regained control in the air, and landed uneventfully 6 minutes later. The official first flight was delayed until a new stabilator could be fitted, and took place on 2 February 1974, again with Oestricher at the controls. The aircraft reached 400 mph and 30,000 feet during the 90-minute flight. The second YF-16 (72-1568) made its maiden flight on 9 May 1974 with Neil Anderson.[16]

The Northrop Competitor

The YF-17 was the culmination of a long line of Northrop designs, beginning with the N-102 Fang in 1956, and continuing through the F-5 series. Northrop's first step toward the F/A-18 was taken in early 1965 with Lee Begin leading the N-300 project. This was essentially an improved F-5E with a stretched fuselage and small leading-edge root extensions (LERXs). The LERX set up vortices over the upper wing surfaces to separate the boundary layer air and improve maneuverability at high angles of attack. The General Electric J85 turbojets were replaced by a pair of 9,000-pound force (lbf) General Electric GE15-J1A1s. A year of wind tunnel testing resulted in the wing being moved from its low position to one

The first YF-17 demonstrator (72-1569) at the Edwards Air Show in November 1974. Not nearly as colorful nor as sleek looking as the YF-16, it was nevertheless a very capable performer. *(Mick Roth)*

The second YF-17 (72-1570) spent most of its Air Force career in a slightly strange looking two-tone camouflage. Surprisingly, the aircraft actually looked better in the camouflage than in the overall silver of the first aircraft. The aircraft was displayed at the November 1975 Edwards Air Show. *(Dennis R. Jenkins)*

high on the fuselage to maximize ordnance flexibility. Interestingly, around the same time Lockheed began a similar project known as the CL-1200 to replace the F-104.[17]

By 1967 the N-300 design had evolved into the P530 powered by a pair of 10,000-lbf GE15J1A2 turbojets, although these were eventually replaced by 13,000-lbf J1A5 versions. This engine used a scaled-down core from the F101 turbofan developed for the B-1A, and had 10 stages of compression, which generated a pressure ratio of 25:1. However, the GE15 had a very modest 0.25 bypass ratio, and the airflow did little more than cool the rear casing of the engine. For this reason, the GE15 was often referred to as a "leaky turbojet" instead of a true turbofan. Nevertheless, this configuration brought a significant benefit—because of the bypassed air, the engine bays required little cooling and could be made of lighter, lower-cost materials.

The trapezoidal wing platform was generally similar to that of the F-5, with a sweep at the quarter-chord line of 20° and an unswept trailing edge. The wing was initially shoulder-mounted with 5° of anhedral, but over the next couple of years the wing moved steadily downward on the fuselage until it ended up at the mid position. The 400 square feet of wing area was over double the 186 square feet on the F-5E. The wing used hinged flaps along the leading and trailing edges, with those on the trailing edge stopping a little over half way to the tip to allow conventional ailerons.

The wing was provided with a prominent LERX that tapered into the fuselage on a level with the cockpit. The LERX made it possible to perform poststall maneuvering at angles of attack exceeding 40°, adding about 50 percent to the lift provided by the basic wing. Extending the LERX ahead of the engine inlets had the additional effect of guiding the airflow smoothly into the inlets and presenting the engines with a flow of relatively undisturbed air at high angles of attack. In addition, a long axial slot was cut into each LERX adjacent to the fuselage, preventing a buildup of air ahead of the inlet in supersonic flight. At low speed and high angles of attack, these slots provided an escape for boundary-layer air that scrubbed across the fuselage ahead of the inlet. In 1968, the LERXs were further enlarged, the forward portions continuing ahead as strakes almost to the nose.

The YF-17 shows its clean lines from overhead. The wing planform was obviously inspired by the F-5 series of lightweight fighters. Note the large BLAD slots cut into the LEX. *(Northrop via the Terry Panopalis Collection)*

The engines were fed by long ducts that admitted air from oval-shaped inlets, which were originally provided with a movable half-cone centerbody, similar to the F-104. By 1971 it had been concluded that Mach 2 performance was not particularly important, and the centerbodies were eliminated. At about the same time, the inlets were made shorter and positioned farther back under the LERX. This made the LERX look like the head of a cobra, a name subsequently adopted by Northrop for the P530. The inlets were further refined throughout 1971, the final shape being a canted D with a slightly rounded leading edge. The top of the inlet was located 4 inches below the underside of the LERX, and the inlets were separated from the fuselage by a large rectangular splitter plate to prevent boundary air from reaching the engines.

The empennage of the P530 was originally fairly conventional, with all-moving stabilators mounted below midlevel and a single vertical stabilizer. Further analysis showed that the P530's ability to fly at extreme angles of attack made the single vertical inadequate since it was often blanketed in the wake of the wing. Twin vertical stabilizers were canted out at almost 45° to remain in free-stream airflow and, to reduce roll cross-coupling, the rudders extended only halfway up the verticals. In 1969, the verticals were doubled in size and moved forward to a position partially overlapping the wing. By late 1970 the vertical surfaces had been further enlarged and their outward cant had been reduced to only 18°. At the same time, the stabilators were enlarged and moved farther aft on the fuselage sides. This long evolution produced a slightly odd appearance, but provided an extremely agile aircraft in most flight regimes.

The resulting aircraft exhibited relaxed longitudinal stability,[18] with a tendency to pitch nose up, enhancing air-combat maneuverability to the point where the limiting factor would be the pilot. However, Northrop did not feel that 1965-vintage fly-by-wire control systems were sufficiently reliable, and retained conventional mechanically signaled flight controls.

The maximum takeoff weight of the P530 was estimated at 40,600 pounds, and a top speed of Mach 2 was expected. The armament was a single M61 20-mm cannon mounted on the centerline underneath the nose, and an AIM-9 Sidewinder could be carried on each wingtip. An array of air-to-ground weapons could be carried on external stations, three under each wing and one on the fuselage centerline.

After the Navy committed to building the F-18, the YF-17 was repainted into Navy colors, although it still displayed a small cobra on the nose in defiance of the Hornet name. Note the lack of a built-in boarding ladder, something typical of Air Force aircraft. *(Dave Begy via the Mick Roth Collection)*

On 28 January 1971, Northrop unveiled the P530 Cobra to the world. The world was apparently unimpressed, and no orders were forthcoming. The mock-up was displayed at the Paris air show, but still there were no takers.

A fact of life is that very few air forces are willing to purchase a foreign aircraft that is not in service with the nation that developed it. The F-5 had been a rare exception, largely because the U.S. government had underwritten much of its procurement cost as foreign aid. Initially the Department of Defense agreed to back the P530 program by partially funding its sales effort in Europe. The reason was simple—the United States knew the Europeans needed a new fighter to replace the F-5 and F-104, and there was nothing upcoming for the U.S. inventory that the European nations could reasonably afford. But the funding had some restrictions upon its use that Northrop later found they could not accept, and the money was never used. This lack of official U.S. backing severely hampered Northrop's efforts to sell the aircraft. Lockheed ran into similar problems.

Northrop tried to interest the Air Force in various versions of the P530, but the Air Force was not in a position to fund either the P530 or Lockheed's CL-1200 program. Nevertheless, the Air Force did find $10 million to fund General Electric to continue development of the GE15 engine under the YJ101 designation.[19]

When the LWF program was announced in 1971, Northrop modified the P530 design into the P600 that would become the YF-17A. However, the P530 had been envisioned as a multirole aircraft with a significant air-to-ground capability, whereas the P600 was to be purely an air-to-air demonstrator. The M61 20-mm cannon was relocated to the upper part of the nose instead of underneath, the position optimized for aerial combat instead of strafing.

The YF-17A was powered by a pair of 14,400-lbf General Electric YJ101-GE-100 turbofans, a further development of the GE15 engine that had been proposed for the P530. The two engines were mounted close together to minimize asymmetric effects in the event of an engine loss. The maximum takeoff weight of the aircraft was initially estimated at 21,000

The first YF-16 flies formation with the first YF-17. The red, white, and blue paint scheme applied to the General Dynamics aircraft certainly attracted more attention, but in reality the two aircraft were fairly close in terms of performance. In the end, both designs enjoyed considerable sales success, although not necessarily in their intended roles. *(Lockheed Martin Tactical Air Systems via Mike Moore)*

pounds, but it soon grew to 23,000 pounds, even though over 900 pounds of its structure was made of graphite/epoxy composite.[20]

The first YF-17A (72-1569) was rolled out at Hawthorne on 4 April 1974, and was trucked overland to Edwards AFB. The program was delayed slightly by problems with the flight certification of the YJ101 engines. The aircraft finally made a 61-minute maiden flight on 9 June 1974 with Hank Chouteau at the controls. On 11 June the YF-17 became the first American fighter to exceed the speed of sound in level flight without the use of an afterburner.[21] The second YF-17A (72-1570) flew for the first time on 21 August 1974, again with Chouteau piloting. The two prototypes carried out a series of 288 test flights totaling 345.5 hours, including 13 hours at supersonic speeds. Late in the test program, a sustained angle of attack of 63° was achieved at 50 knots, validating the extensive work accomplished on this part of the flight regime.[22]

The Fly-off

The fly-off between the YF-16 and the YF-17 began as soon both contractors completed preliminary flightworthiness demonstrations. There was an attempt to get as many pilots as possible to fly both the YF-16 and YF-17 to aid in the overall evaluation, although it is unclear exactly how many actually managed to fly both designs. The prototypes never flew against each other, but they did fly against all current Air Force fighters as well as against MiG-17s and MiG-21s that had been "acquired" by the United States.

In the meantime, the governments of Belgium, Denmark, Norway, and the Netherlands had begun to consider replacing their F-104s. They formed the Multinational Fighter Program Group (MFPG) to choose a successor on the basis of evaluations of the Dassault Mirage F.1, General Dynamics YF-16, Northrop YF-17, and Saab JA37 Viggen.

The first YF-17 in a three-tone blue camouflage with Marine markings on one side and Navy markings on the other. These markings changed substantially while the aircraft was being used as a demonstrator. Note that the insides of the vertical stabilizers are not camouflaged—they would be later. The national insignia would also move to the forward fuselage. *(Northrop via the Terry Panopalis Collection)*

The winner of the LWF contest would probably be the favored candidate, but the MFPG wanted assurance that the United States was going to buy the aircraft for itself before they made a decision.

In September 1974, Secretary of Defense James R. Schlesinger announced that he was considering production of the LWF winner. Up until that time, the LWF program had been largely

The second YF-17 at NAS Miramar with "F-18 Prototype" markings on the rear fuselage. Note the relatively narrow track of the landing gear and the single-wheel nose unit. *(Mick Roth)*

an academic exercise, but the possibility of a large European order led the Pentagon to reconsider. However, the military now wanted a multirole aircraft rather than a simple air combat fighter. This was largely because the F-15 had been designed as a dedicated air-superiority fighter, and a multirole fighter was needed to replace the F-4 and F-105 fighter-bombers.[23]

Interestingly, at the same time the mission requirements were changing away from their lightweight fighter origins, the aircraft was renamed the Air Combat Fighter (ACF). On 11 September 1974 the Air Force announced plans to buy 650 Air Combat Fighters, with the possibility that this could be substantially increased.

At the beginning of the competition it was generally assumed that Northrop would ultimately be the winner. Their years of lightweight fighter experience, the successful T-38 trainer, and the amount of time and money invested in the original P530 project were thought to give them an unbeatable advantage. So it was with some surprise that, on 13 January 1975, Secretary of the Air Force John McLucas announced that the YF-16 had been selected. The YF-16 was a little faster than the YF-17, and its F100 engine provided commonality with the F-15, greatly reducing the logistics tail. The J101 engine was a new, relatively untried powerplant that would require a large investment in tooling, spare parts, and documentation.

The MFPG subsequently ordered versions of the F-16, and the F-16 quickly became the most-produced jet fighter of its generation.

YF-17 Description

Northrop reported that the development of the YF-17 and YJ101 engine consumed over 1,000,000 man-hours and 5,000 hours of wind tunnel testing. The two YF-17s used conventional semimonocoque stressed-skin construction, employing high-strength aluminum alloys as its primary material. Steel and titanium were used in areas that required particularly high strength or had to endure high temperatures for prolonged periods. The fuselage consisted of a stressed skin supported by frames, longerons, and bulkheads. Graphite-epoxy composite was used for the engine bay doors, landing gear doors, and various equipment access doors. The forward fuselage contained the ranging radar, air data computer, battery, and emergency power unit.[24] The pilot sat on a Stencel Aero 3C ejection seat inclined at 18°. A bubble canopy provided excellent visibility in all directions, and a JLM International heads-up display provided all necessary flight and combat information.[25]

Internal fuel was carried in four bladder-type cells located in the midfuselage behind the cockpit. The aircraft had an in-flight refueling receptacle located in the nose ahead of the windscreen, and was fitted for single-point ground refueling. Two 600-gallon drop tanks could be carried on the inboard wing pylons, while the centerline pylon could carry a 300-gallon drop tank.

The thin dry wing consisted of left and right panels attached to the fuselage with shear bolts. The wing used a relatively thick skin over a conventional mutispar structure with machined and welded fuselage attach ribs. Various sections of the wing, including the LERX, trailing edge, trailing-edge flaps, and ailerons, used a Nomex honeycomb core with graphite-epoxy facesheets.

The all-moving stabilators used full-depth honeycomb cores with machine-tapered and -sculpted aluminum skins, aluminum leading-edge wedges, and a full-span machined aluminum channel spar. The vertical stabilizers used a thick skin over a conventional mutispar structure. The leading and trailing edges used aluminum honeycomb core with a laminated graphite-epoxy spar and graphite-epoxy facesheets.

Mechanical power was supplied by two fully independent 3,000-psi hydraulic systems. The left engine drove one system, which provided half the power for the flight controls and the leading-edge flap actuators, and all of the power for the landing gear, brakes, cannon, and pitch control augmentation system. The right engine drove the second system, which

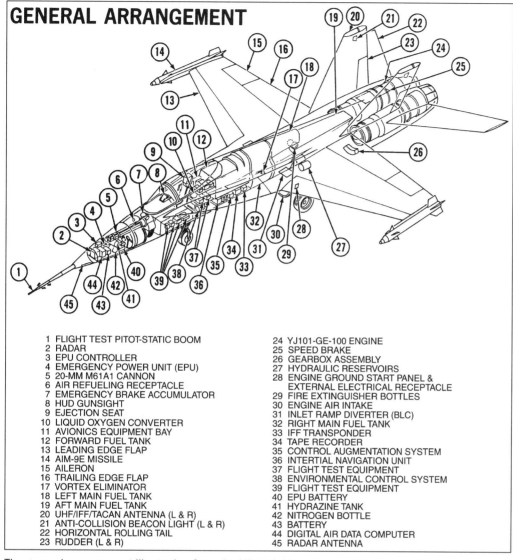

GENERAL ARRANGEMENT

1 FLIGHT TEST PITOT-STATIC BOOM
2 RADAR
3 EPU CONTROLLER
4 EMERGENCY POWER UNIT (EPU)
5 20-MM M61A1 CANNON
6 AIR REFUELING RECEPTACLE
7 EMERGENCY BRAKE ACCUMULATOR
8 HUD GUNSIGHT
9 EJECTION SEAT
10 LIQUID OXYGEN CONVERTER
11 AVIONICS EQUIPMENT BAY
12 FORWARD FUEL TANK
13 LEADING EDGE FLAP
14 AIM-9E MISSILE
15 AILERON
16 TRAILING EDGE FLAP
17 VORTEX ELIMINATOR
18 LEFT MAIN FUEL TANK
19 AFT MAIN FUEL TANK
20 UHF/IFF/TACAN ANTENNA (L & R)
21 ANTI-COLLISION BEACON LIGHT (L & R)
22 HORIZONTAL ROLLING TAIL
23 RUDDER (L & R)

24 YJ101-GE-100 ENGINE
25 SPEED BRAKE
26 GEARBOX ASSEMBLY
27 HYDRAULIC RESERVOIRS
28 ENGINE GROUND START PANEL &
    EXTERNAL ELECTRICAL RECEPTACLE
29 FIRE EXTINGUISHER BOTTLES
30 ENGINE AIR INTAKE
31 INLET RAMP DIVERTER (BLC)
32 RIGHT MAIN FUEL TANK
33 IFF TRANSPONDER
34 TAPE RECORDER
35 CONTROL AUGMENTATION SYSTEM
36 INTERTIAL NAVIGATION UNIT
37 FLIGHT TEST EQUIPMENT
38 ENVIRONMENTAL CONTROL SYSTEM
39 FLIGHT TEST EQUIPMENT
40 EPU BATTERY
41 HYDRAZINE TANK
42 NITROGEN BOTTLE
43 BATTERY
44 DIGITAL AIR DATA COMPUTER
45 RADAR ANTENNA

The general arrangement illustration from the YF-17 flight manual shows the location of the major systems and components. The prototypes were not meant to be truly representative of an operational aircraft, although Northrop did include sufficient space to accommodate most operational systems. *(Northrop)*

provided the other half of the power for the flight controls and leading-edge flap actuators, as well as all of the power for the trailing-edge flaps, nosewheel steering, and emergency main gear extension system.

The primary flight controls consisted of ailerons, rudders, and the stabilators. Basic roll control was provided by the ailerons, supplemented by differential deflection of the stabilators. All primary flight controls were powered by redundant hydromechanical actuators and normally commanded through the electric control augmentation system (CAS). Responses from sensors within the system were used by a digital air data computer (DADC) to provide relatively constant stick forces throughout the flight envelope. A graphite-epoxy speedbrake was located above and between the engines and was positioned by a single hydraulic actuator.

The YJ101-GE-100 engines were installed using two thrust mounts and an aft steady-rest for each engine. The engines were removed by lowering them vertically from the engine bay without disconnecting any of the empennage controls. The accessory drive system

NORTHROP LIGHTWEIGHT FIGHTER EVOLUTION					LWF PROPOSAL	CARRIER-BASED F-18A
HIGH-WING FORWARD INLETS	LARGER LERX INLETS UNDER LERX	TWIN VERTICAL STABILIZERS LARGER LERX	CONTOURED LERX EXTENSION LARGER VERTICAL STABILIZERS	REFINED FUSELAGE SHORTER INLETS	P610 TWIN ENGINE LWF / P600 SINGLE ENGINE LWF	YF-17 PROTOTYPE / LAND-BASED F-18L
N-300	P530	P530-1	P530-2	P530-3	P6x0	
1966	1967	1968	1969	1970	1971-72	1973 1976

The evolution of the Northrop lightweight fighter from 1966 to 1976. *(Northrop)*

consisted of two airframe-mounted, engine-driven gearboxes interconnected by a common starter gearbox. During engine removal, the gearboxes remained in the aircraft, while the drive shaft was mechanically disconnected to avoid disturbing aircraft systems.

Armament consisted of a single M61 20-mm cannon and two AIM-9 Sidewinder missiles. The cannon was located in the forward fuselage and was pallet-mounted for ease of maintenance. Additional ordnance could be carried on four wing stations and a single centerline station.

The aircraft had an advertised empty weight of 17,180 pounds, plus 90 pounds of unusable fuel. The basic mission takeoff weight (full internal fuel, full 20-mm ammunition, two AIM-9s) was 24,580 pounds. The maximum takeoff weight was 34,280 pounds.[26]

During the LWF competition, the YF-17 demonstrated a top speed of Mach 1.95, a peak load factor of +9.4g, a maximum altitude of just over 50,000 feet, and a sea-level rate of climb of over 50,000 feet per minute. The aircraft could achieve a controllable angle of attack of 34° in level flight, and 63° during a zoom climb. Control could be maintained at indicated airspeeds as low as 20 knots at high angles of attack. Northrop claimed that the aircraft had no angle-of-attack limitations, no control limitations, and no adverse departure tendencies.[27]

A Second Chance: The F/A-18A

Losing the LWF/ACF competition might have been the end of the line for the Northrop design, were it not for the U.S. Navy's desire for a new fighter. Some Navy officers had quietly sided with the Fighter Mafia and were expressing interest in a low-cost alternative to the Grumman F-14 Tomcat. Also, the Marine Corps was increasingly worried that the F-14 would not be developed into the true multirole aircraft they needed. Unlike the Air Force, senior Navy officials were more open to the idea. The Naval Fighter-Attack, Experimental (VFAX) program was initiated to procure a multirole aircraft that would eventually replace the F-4 Phantom II, A-4 Skyhawk, and A-7 Corsair II. Perhaps part of the reason the VFAX did not run into the same opposition that the LWF had was that the senior officials made it clear from the beginning that the aircraft was intended to complement the F-14, not replace it. The Navy was solidly behind the F-14 as its fleet defense fighter, even if cost overruns meant it would have to procure fewer of them in the long run.[28]

During the summer of 1973, Secretary of Defense Schlesinger ordered the Navy to evaluate the Air Force lightweight fighters. This was reinforced in August 1973 when Congress mandated that the Navy pursue a lower-cost alternative to the F-14 since cost overruns, plus financial instability at Grumman, had finally reached a critical level. A stripped version of the Tomcat (the F-14X) had already been proposed by Grumman, but was summarily rejected since it greatly reduced the aircraft's capabilities, but only slightly reduced its actual cost. Similarly, a "navalized" F-15 was proposed by McDonnell Douglas, but ended up nearly as heavy and expensive as the F-14 it was supposed to supplement.

On 10 May 1974 the House Armed Services Committee withheld the Navy's $34 million request to continue the VFAX program. Apparently having forgotten the sorry experience with the F-111, Congress wanted the Air Force and Navy to purchase similar aircraft based on the LWF candidates. However, the Navy (unlike the Air Force, at the time anyway) wanted the VFAX to be capable of filling both air-to-air and air-to-ground roles, something the LWF demonstrators were not designed to do.[29]

In August 1974 Congress took the $34 million originally sought for the VFAX and diverted it to a new program known as the Navy Air Combat Fighter (NACF). Congress directed that the NACF make maximum use of the technology and hardware developed for the Air Force's LWF/ACF. Naval aviators began evaluating the YF-16 and YF-17 at Edwards AFB, and both contractors assured the Navy that a carrier-based version of their respective aircraft was possible. At the same time, both General Dynamics and Northrop realized they had never built a carrier-based fighter, and that this increased risk in the eyes of the government. In order to mitigate this risk, both companies actively sought suitable partners.

On 27 September 1974 General Dynamics announced it would team with LTV (also located in Dallas/Fort Worth) to propose an NACF based on the YF-16. The Navy version would include a BVR radar and expanded multirole capabilities. LTV (Vought) brought substantial Navy experience, most recently with the F-8 Crusader and A-7 Corsair II.

Northrop also felt they had a potential candidate for the NACF in the YF-17, since the design seemed to have greater potential for growth into a radar-equipped multirole aircraft.[30] To address their lack of carrier experience, Northrop teamed with McDonnell Douglas on the naval adaptation of the YF-17. The navalized P630 was extensively modified, and McDonnell Douglas called the resultant aircraft the Model 267.

And the Winner Is . . .

When the YF-16 was announced as the winner of the LWF/ACF competition, the Navy's response was overwhelmingly negative. The Navy did not believe that a single-engine aircraft with narrow-track landing gear could be adapted to carrier operations without a great deal of effort, and informed the Pentagon it had no intention of procuring an F-16 derivative. Their rationale was that the changes required to make the F-16 suitable for carrier operations would essentially result in a new aircraft, eliminating the perceived benefits of a common design. During the next four months, the Navy negotiated with Congress and the Secretary of Defense to develop a navalized derivative of the losing YF-17 design.[31]

On 2 May 1975 the Navy announced it had received approval to develop an aircraft based on the YF-17. But many of the arguments against the YF-16 also applied to the YF-17. In effect, the Navy was asking Northrop and McDonnell Douglas to take the basic configuration of the YF-17 and design a new aircraft around it. In a rare bout of bureaucratic honesty, the Navy redesignated the aircraft F-18 in recognition of the substantial differences.[32] The final F-18A would not share a single essential dimension or primary structure with the YF-17 demonstrators.

The first YF-17 was retained by Northrop as a demonstrator for the proposed F-18L, while the second YF-17 was turned over to the Navy for pilot familiarization and public relations tours. It subsequently flew from the Pacific Missile Test Center (PMTC) at Point Mugu, California, the Naval Air Test Center (NATC) at Patuxent River, Maryland, and the Naval Weapons Center (NWC) at China Lake, California. Both aircraft also flew some test flights for the National Aeronautics and Space Administration (NASA).

According to the original plan, the aircraft was intended to be procured in three closely related models—the single-seat F-18, which would replace the F-4 in the fighter role, the single-seat A-18, which would replace the A-4 and A-7 in the attack role, and the two-seat TF-18 trainer. The F-18 and the A-18 were to share the same basic airframe and engines, but were to use different avionics and have different weapons capabilities. The two-seat TF-18 was to retain the full mission capability and armament of the F-18, but would have a slightly reduced fuel capacity. On 1 March 1977 Secretary of the Navy W. Graham Claytor announced that the F-18 would be named "Hornet."

The F-18 program went ahead with the award of letter contracts in November 1975 to General Electric for the development of the F404 turbofan and on 22 January 1976 to McDonnell Douglas for 9 single-seat and 2 two-seat full-scale development (FSD) aircraft.

This was the first F-18A artist concept released. Note that the rudders extend only halfway up the verticals. It is interesting to note that the artist angled the wingtip launch rails—in the wrong direction! Because of a built-in wing twist, the production rails are angled down somewhat so that they are parallel to the direction of travel. *(Frederick A. Johnsen Collection)*

McDonnell Douglas and Northrop agreed that production of the baseline F-18 would be split roughly 60/40 between the two contractors, with the extra 20 percent largely representing final assembly. McDonnell Douglas was the "prime" contractor for the naval versions. In the event of orders being received for the F-18L land-based version, these proportions would be reversed, with Northrop performing final assembly and being the prime contractor. Regardless of variant, Northrop was to build the center and aft fuselage sections, as well as both vertical stabilizers, while McDonnell Douglas would build the wings, stabilators, and the forward fuselage and cockpit.[33]

Baseline requirements called for 780 Hornets to equip Navy fighter and attack squadrons (VF and VA, respectively), as well as Marine fighter-attack squadrons (VFMA). As it turned out, careful attention to some design details made it possible to merge the fighter and attack versions into a single aircraft. The ability to develop a single variant was made possible partly by careful redesign of the two stores stations located on the lower corners of the air intakes. In the fighter role, these stations would carry AIM-7 Sparrow air-to-air missiles, while in the attack role they would carry a forward-looking infrared (FLIR) pod on the left side and a laser spot tracker/strike camera (LST/SCAM) pod on the right side. Constant improvements in radar and processing capability meant a single avionics suite could perform both the air-to-air and air-to-ground missions. Multifunction displays allowed a common cockpit to be developed, eliminating the last major difference between the designs. This aircraft began to be referred to as F/A-18A in Department of Defense press releases sometime during 1980, but the designation did not become official until 1 April 1984. The combat-capable two-seat trainer was successively redesignated TF-18A and then F/A-18B.[34]

Growing the YF-17

The McDonnell Douglas Model 267 retained the overall configuration of the Northrop YF-17 lightweight fighter prototype, with its two engines, twin outward-canted vertical stabilizers, and prominent LERX. However, both the airframe and the undercarriage were significantly strengthened for carrier operations, and folding wings, catapult attachments, and a strengthened tail hook were added. In addition, the fuel capacity was increased by 4,460 pounds to meet the Navy-specified mission radius and safety reserves.

The prominent LERX from the YF-17 was retained, but renamed a leading-edge extension (LEX). Compared to the YF-17, the wing was increased from 350 to 400 square feet, with increases in both span and chord in order to improve the low-speed performance. The trapezoidal platform (swept on the forward edges but straight on the trailing edges) incorporating full-span leading-edge flaps and hydraulically actuated single-slotted flaps on the inboard trailing edges. These surfaces were all under computer control, setting the most desirable angle to give optimal performance throughout the performance envelope. The deployment angles of the trailing-edge flaps were increased from 30° to 45°. A 96-gallon integral fuel-cell was added in each wing, but most of the internal fuel continued to be carried in the fuselage. A "snag" was added to the leading edge of both the wings and stabilators to generate a vortex in order to prevent a slight flutter that had been discovered on the F-15 stabilator.[35]

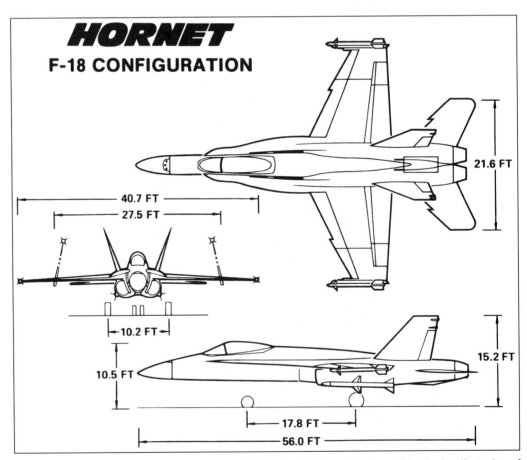

The first three-view drawing released of the F-18A clearly shows the "snag" in the leading edge of the wing and horizontal stabilator, although the stabilator is not shown as installed on the full-scale development (FSD) aircraft. Also note that the ailerons do not extend all the way to the wingtips, and that the wing fold is not along the flap/aileron break. *(Boeing)*

The F/A-18 uses five pairs of aerodynamic control surfaces—ailerons, leading-edge flaps, trailing-edge flaps, stabilators, and rudders. Pitch control is provided by the collective operation of the stabilators and symmetric leading- and trailing-edge flaps. Originally, roll control used differential stabilator and aileron movement, although asymmetric leading- and trailing-edge flap movement was added later to improve roll characteristics. Directional control is provided by symmetric rudder deflection, while a later change added rudder tow-in/flare-out to increase pitch control in the landing/takeoff configuration. A twin-hinged hydraulically activated speedbrake is mounted on the upper rear fuselage, between the vertical stabilizers, providing minimal pitch change when the speedbrake is extended.

The twin vertical stabilizers are canted outboard at approximately 20° to keep them in relatively clean air at high angles of attack. The verticals are mounted far forward in order to close the aerodynamic gap between the trailing edge of the wing and the leading edge of the vertical stabilizer, resulting in a smooth and low-drag airflow. The forward position of the verticals also reduced airflow interference around the engine nozzles. Compared to the YF-17, the stabilators are larger and have a lower aspect ratio.

The width of the aft fuselage was increased by 4 inches over the YF-17, the engines canted outward at the front, and the fuselage spine made significantly wider and taller. The spine contains four main fuel tanks (containing 426, 249, 200, and 530 gallons, front to rear), and all tanks and fuel lines are self-sealing, with foam in the wing tanks. There is a retractable refueling probe on the right side of the fuselage just ahead of the cockpit, but no provisions exist for an Air Force–style (boom) in-flight refueling receptacle.

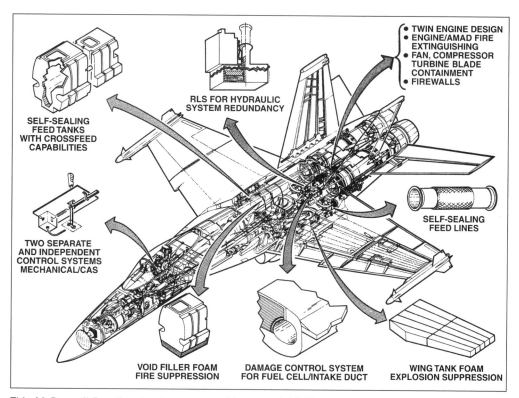

This McDonnell Douglas drawing appeared in several 1977 periodicals touting the advanced survivability features of the F-18A. This was an important selling point at the time since lessons learned in Vietnam had showed that many earlier American fighters had critical single-point failure modes. *(Boeing)*

Since the Navy wanted an all-weather capability and the ability to use AIM-7 Sparrow radar-homing missiles, the small radar of the YF-17 was replaced with a more powerful installation. At the end of 1977, the Hughes (now Raytheon) AN/APG-65 digital multimode pulse-Doppler radar was selected over its Westinghouse (now Northrop Grumman) AN/APG-66 competitor. The shape of the nose was altered in order to accommodate the 28-inch radar antenna necessary to meet the Navy's 30 nautical mile search requirement.[36]

The undercarriage of the YF-17 had a track of 6 feet 10.75 inches, but McDonnell Douglas increased this to 10 feet 2.5 inches for greater stability during carrier landings. The totally redesigned landing gear now met the 24 feet per second descent rate requirement that is needed for arrested carrier operations. The main undercarriage units retract aft and rotate ("plane") through 90° to lie flat underneath the air intake ducts. A twin-wheel nose gear retracts forward into the nose and incorporates a launch bar for catapult operations.

The engine air intakes are set well back underneath the LEXs, protecting the engine intakes somewhat from disruption of the airflow caused by high-angle-of-attack flight. Since there was no requirement for the F/A-18 to exceed Mach 2, the aircraft retains the simple fixed-geometry inlets developed for the YF-17. The two-dimensional D-shaped intakes have a fixed splitter plate mounted next to the fuselage, and ducts cut into the top of the LEX permit bleed air to be ejected upward into the airflow generated by the LEX. The intake ramps and boundary layer splitter plates are solid at the front end, with perforations directly ahead of the inlet to permit sluggish boundary layer air to be bled away and dumped via spill ducts on top of the LEX.

All of these changes added over 10,000 pounds to the YF-17's 23,000-pound gross weight. Approximately half of the structural weight of the F/A-18A is made up of aluminum, while steel contributes about 16.7 percent. Titanium makes up about 12.9 percent, being used for significant portions of the wings, vertical stabilizer, and stabilator attachments, and the wing-fold hinges. About 40 percent of the aircraft's surface area is covered by graphite-epoxy composite skin, making up 9.9 percent of the aircraft's weight. The remaining structural weight is accounted for by various other materials such as plastic and rubber.

The YJ101 turbofans that powered the YF-17 were replaced by their F404-GE-400 derivatives, rated at 15,800 lbf with afterburner. The F404 has a bypass ratio of 0.34, making it more of a true turbofan rather than a "leaky" turbojet. The engine has essentially the same thrust as the J79 turbojet, but weighs only half as much, is 25 percent shorter, and has 7,700 fewer parts. Compared to other recent turbofans, the F404 experienced relatively few developmental problems and is extremely resistant to compressor stalls even at high angles of attack. The engine is remarkably responsive, being able to accelerate from idle to full afterburner in only 4 seconds.

The Hornet has nine external stores stations—one at each wingtip, two underneath each wing, one on each corner of the fuselage just aft of the air intakes, and a centerline station under the fuselage. The wingtip stations can carry only AIM-9 Sidewinders and instrumentation pods. The corner stations behind the air intakes carry AIM-7 Sparrows when the Hornet is operating in the fighter role, or Ford Aerospace[37] AN/AAS-38 NITE[38] Hawk FLIR and Martin Marietta[39] AN/ASQ-173 LDT/SCAM pods when operating in the attack role. The ASQ-173 did not include a laser designator, so initially the Hornet had to rely upon other aircraft (usually an A-6E or OV-10) or ground forces to designate targets for its laser-guided bombs. The M61A1 20-mm cannon was retained in the forward fuselage.

The F/A-18 incorporates a quadruple-redundant digital fly-by-wire (FBW) flight control system, the first of its kind to be installed in a production aircraft.[40] Stick and rudder inputs are directed to a computer, which interprets them and issues the appropriate commands to the hydraulic actuators that operate the various control surfaces.[41] The system operates by the principle of majority vote—if one of the four systems disagrees with the other three, this is interpreted as a failure, and the dissenting system is ordered to shut down. Redundancy is

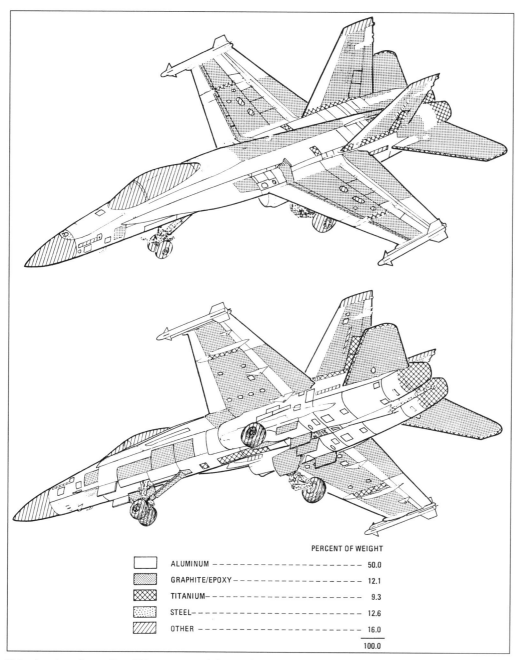

		PERCENT OF WEIGHT
	ALUMINUM	50.0
	GRAPHITE/EPOXY	12.1
	TITANIUM	9.3
	STEEL	12.6
	OTHER	16.0
		100.0

This drawing shows the different materials used to manufacture the F-18A. The drawing predates the production aircraft, so the final percentages do not represent most aircraft. *(U.S. Navy)*

such that should a second system fail, the remaining two systems can still operate the controls so long as they remain in agreement. In the unlikely event of all four systems failing, there is a direct electrical link (DEL) to the control surfaces. In reality, there are multiple DEL modes, depending upon which part of the normal flight control system has failed. DEL commands are still processed by the flight control computers and, therefore, DEL is still a fly-by-wire system. However, in DEL mode the normal stability augmentation is absent, and the aircraft exhibits degraded handling qualities. There is also a direct mechanical backup for the stabilators that totally bypasses the flight control computers and gives the pilot some degree of pitch and roll control in an extreme emergency. Originally the control laws did not incorporate any provisions to prevent the pilot from overstressing the airframe, but during the FSD program this logic was added in conjunction with the center-of-gravity control system.

The Hornet underwent extensive wind tunnel evaluations involving full-scale models as well as the small spin-test model shown here in the NASA-Langley spin tunnel. *(NASA)*

The Hornet had always been intended as a single-seater, so a great deal of attention was paid to reducing the pilot workload through the extensive use of automation, and the Hornet was the first fighter with a "glass cockpit." Many of the conventional dial-type instruments were eliminated and their information displayed on monochrome multi-function displays (MFD). The pilot was provided with a hands-on throttle and stick (HOTAS), with all the controls required for combat being located on either the throttle lever or control stick for easy access. Unlike the contemporary F-16, the F/A-18 used a conventional center control stick, largely because a side stick could not provide the deflection to operate the emergency mechanical backup to the stabilator. A Martin Baker Mk 10 (SJU-5/6) zero-zero rocket-assisted ejection seat was provided for the pilot (or pilots, in the case of the two-seater), but the 18° reclined angle was much less than used in the F-16.

The Doomed F-18L

While the carrier-based F-18A was being designed by McDonnell Douglas, a land-based F-18L was being designed by Northrop. Since it did not have to carry any equipment for carrier operations, the F-18L was expected to be significantly lighter and better-performing than the carrier-based version. It was anticipated that the F-18L would be an attractive proposition for those foreign air forces that could afford an aircraft with greater capabilities than those of the F-5 and F-104. Export prospects were dimmed somewhat by the overwhelming foreign orders being enjoyed by the F-16, but F-18L proponents pointed out the advantages of two engines and a more sophisticated avionics suite.[42]

The F-18L never progressed beyond the mock-up stage. Note the third weapons station under the wing. At various times the mock-up was configured as both a single-seater and a two-seater (shown). Surprisingly, the markings on the side of the fuselage say "F/A-18L." *(Northrop via Don Logan)*

The F-18L had a normal operating weight that was 7,700 pounds less than the baseline F-18A. Of this, 3,500 pounds was attributed to a reduced fuel load, while the rest was structural weight saved by deleting naval equipment.[43] The aircraft had lighter-weight landing gear, nonfolding wings, reduced longeron and frame thickness in some areas, lighter arresting gear, and a simplified avionics package. The most obvious external difference between the F-18A and F-18L was the elimination of the wing and stabilator snags.[44] The two designs shared 71 percent of their components by weight, and 90 percent of the high-value systems.[45]

The same APG-65 radar and electronic countermeasures (ECM) suite used in the F-18A were standard equipment in the F-18L, although Northrop offered several alternative ECM suites (mainly using Air Force equipment instead of Navy) based on customer requirements. Interestingly, the F-18L carried only 400 rounds of 20-mm ammunition instead of the 578 carried by the F-18A. As early as 1978 Northrop envisioned a nose-mounted sensor pallet that could replace the M61 20-mm cannon—the same basic concept would be implemented on the F/A-18D(RC) much later (the suffix RC represents *reconnaissance-capable*). A Martin Baker US10V ejection seat was used instead of the naval advanced-concept ejection seat (NACES) version used by the Navy, the primary difference being a lack of sea survival equipment. Northrop also proposed a TF-18L two-seat version with identical fuel capacity (something the F-18B/D lacked) and the same overall length.[46]

Internal fuel was carried entirely in fuselage tanks, as on the YF-17, although the internal wing tanks developed for the F-18A were available as an option. One major difference between the F-18L and the F-18A was the positioning of the external stores stations. The

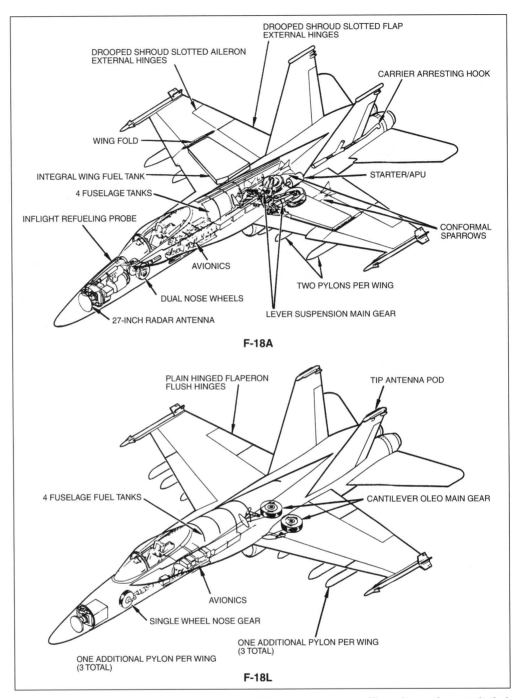

The major differences between the F-18A and F-18L are shown here. All carrier equipment, including folding wings, tail hook, and nose gear launch bar, were deleted. The landing gear was much simpler, looking more like the YF-17's than the F-18's. Although the baseline F-18L used the same basic avionics as the F-18A, Northrop offered Air Force electronic countermeasures (ECM) systems and radios for those customers that wanted commonality with their existing aircraft. *(Northrop)*

Northrop aircraft did not include the two air intake corner stations, but instead featured three pylons under each wing, in addition to the wingtip AIM-9 stations. The air intake stations could be added as an option, carrying the same AIM-7s or sensor pods as the F-18A. The external fuel tanks offered by Northrop included 300-gallon, 450-gallon, and 610-gallon units. The aircraft was rated at +9/−3g through most of its flight envelope, although the air intake and centerline fuselage stations could not support loads above +7.5*g*. The airframe was designed for 6,000 flight hours during 5,000 flights.[47]

Trouble in Paradise

As part of the original partnership arrangement, it had been agreed that McDonnell Douglas would be the prime contractor for the carrier-based F-18A, with Northrop being the prime contractor for the F-18L. But the partnership between the two aerospace giants began to show strains almost immediately. In particular, a major disagreement arose over sales of the F-18L. Northrop felt that whenever foreign purchasers showed an interest in acquiring the F-18L, McDonnell Douglas would mount an active sales effort, putting the F-18A in direct competition. Northrop management became very disturbed about what it perceived to be McDonnell Douglas's violation of the terms of their agreement.

In October 1979, a series of lawsuits filed by Northrop claimed that McDonnell was unfairly using Northrop technology developed for the F-18L to sell its own F-18A abroad. In particular, Northrop charged that McDonnell Douglas was trying to sell a version of the F-18A to Israel that competed directly with the F-18L. Northrop asked the courts to restrain McDonnell Douglas from selling any version of the F-18 that took advantage of Northrop technology to the detriment of the latter company. The case dragged through the courts, and was not settled until April 1985 when it was agreed that McDonnell Douglas would be prime contractor for all existing and future versions of the Hornet. McDonnell Douglas agreed to pay Northrop $50 million for the remaining rights to the original design, without admitting any fault in the case. It was largely a moot point by then since Northrop had already terminated all work on the F-18L, and most export orders had been captured by the F-16.[48]

In retrospect, part of the F-18L's problems can be attributed to the fact that few foreign military sales successes include aircraft with no history of demonstrated U.S. military service. There have been some exceptions to this rule, but not many. The most notable was the Northrop F-5 series, which sold over 2,700 aircraft to 30 countries. Perhaps an overoptimistic Northrop truly believed it could duplicate its F-5 triumph with the F-18L.

F/A-18A

The first FSD F-18A [Bureau Number (BuNo) 160775] taxied out during its official rollout ceremonies in St. Louis on 13 September 1978. An overall white-and-blue paint scheme with gold trim had "NAVY" on the left side of the fuselage and "MARINE CORPS" on the right. The aircraft's maiden flight took place at Lambert Field on 18 November 1978 with Jack E. Krings at the controls. Krings found the F-18 to be remarkably stable and easy to handle, and the 50-minute first flight was uneventful, reaching a maximum speed of 300 knots and an altitude of 24,000 feet.[49] By early December 1978 the aircraft had gone supersonic, and less than a year later, on 18 September 1979, made its hundredth flight.[50]

Most flight testing was conducted at the Naval Air Test Center at Patuxent[51] River, where nine F-18As and two TF-18As underwent an intensive flight test program, aided by a modified T-39 radar test platform. Previously, new aircraft had been tested at a variety of locations, usually near the home plant of the major subsystem manufacturers. For the F-18, the Navy developed the "principal site concept" that concentrated almost all F-18 testing at Pax River. Navy pilots also were involved earlier in the process than normal, flying many missions that previously would have been conducted by company test pilots.[52] The first Navy pilot to fly the Hornet was Lieutenant Commander John Padgett in March 1979 during Navy Preliminary Evaluation period 1 (NPE-1).[53]

Several structural test articles were constructed and used for various purposes. One was specifically dedicated to landing impact testing to make sure the new design could withstand the 24 feet per second descent rate required by the Navy. This article was used for airframe flex testing. *(Frederick A. Johnsen Collection)*

As with most flight test programs, each prototype had specific duties. The first aircraft (BuNo 160775) opened the flight envelope and investigated flutter characteristics. The next aircraft (BuNo 160776) was assigned to engine evaluations and performance testing. Carrier qualifications began on 30 October 1979 with the third FSD aircraft (BuNo 160777) aboard the USS *America* (CV-66). The tests lasted only 4 days, but included 17 touch-and-

This T-39D (BuNo 150987) was equipped with an APG-65 radar in its modified nose during initial Hornet testing and training. This photo was taken in May 1989 after the aircraft was assigned to the Test Pilot School. *(Dr. J. G. Handelman via the Terry Panopalis Collection)*

The first FSD F-18A (BuNo 160775) was overall white with royal blue and gold trim. The aircraft carried "NAVY" on the left fuselage, and "MARINES" on the right. The snags on the wing and horizontal stabilator leading edge are visible. *(McDonnell Douglas via the Terry Panopalis Collection)*

go landings and 32 arrested landings during 14 flight hours. In addition, the aircraft was loaded on each of the elevators to check clearances and parked at various locations on the flight and hanger decks, and simulated maintenance was conducted to verify fit and function with existing equipment. Later carrier compatibility trials included the first fully automatic landing at Pax River on 22 January 1982 with Peter Pilcher at the controls. The USS *Carl Vinson* (CVN-70) subsequently hosted a second series of sea trials consisting of 63 catapult launches and traps.

The fourth F-18A (BuNo 160778) was used as a structural flight test aircraft, while the fifth (BuNo 160779) was assigned to avionics and weapons systems development. The original flight test program called for the sixth aircraft (BuNo 160780) to investigate high angles

The fourth FSD F-18A (BuNo 160778) was used as a structural flight test aircraft. Tests included measuring the stresses associated with dropping large weapons loads, such as the nine 1,000-pound Mk 83 bombs shown here. In this case two bombs were carried on each wing station, and a single bomb was carried on the fuselage centerline station. *(Frederick A. Johnsen Collection)*

The fifth FSD F-18A (BuNo 160779) shows the snags in the leading edge on the wing and horizontal stabilator. This aircraft was painted gloss white with royal blue trim and was assigned to avionics and weapons system testing. *(Tony Landis Collection)*

of attack, departure resistance, and degraded-flight-control-mode flying qualities with no plans for intentional departures or spins. However, as a safety measure the aircraft was configured with an emergency recovery system (ERS) on the upper aft fuselage. The ERS included a spin chute and backup electrical, hydraulic, and fuel boost systems to facilitate recovery from an out-of-control situation and/or dual engine flame-out. Subsequent to the loss of an aircraft in late 1980 due to what appeared to be an unusual spin (very low yaw rate), the flight test program was restructured to investigate spin recovery characteristics. Even at that time, program personnel were careful to avoid reference to a "spin program," referring instead to a "spin recovery program." Of course, investigation of spin recovery characteristics requires that one first intentionally enter a spin.

The first TF-18A (BuNo 160781) and the seventh F-18A (BuNo 160782) were both assigned to weapons systems testing, including stores release demonstrations. The eighth single-seater (BuNo 160783) was used for general performance testing, and the second TF-18A (BuNo 160784) performed accelerated engine service testing to validate the long-term maintenance needs of the F404 engine. The last FSD F-18A (BuNo 160785) was used to validate maintenance procedures.[54]

Weapons systems testing proceeded quickly. The first live AIM-9 was launched by Bill Lowe in December 1979, scoring a near-miss on a BQM-34 target drone (it would have destroyed the drone if a live warhead had been fitted). Three months later, Lowe launched the Hornet's first AIM-7, scoring a direct hit on an unspecified drone.[55] The placement of the M61A1 20-mm cannon directly ahead of the cockpit had caused some concern about hot gun propellant particles pitting the windscreen. Various tests were conducted, including one when the entire 578-round magazine was fired in a single continuous burst. An examination of gun gas patterns showed that they did not pose a pit-

The last FSD F-18A (BuNo 160785) was finished to near-production standard. Note the early configuration of the ECM fairing on the trailing edge of the vertical stabilizer. *(Frederick A. Johnsen Collection)*

ting problem for the windscreen, a visibility problem for the pilot, or an ingestion problem for the engines.

During the weapons trials a minor flutter problem was uncovered with the underwing pylons, solved by moving the stores rack 5 inches forward. Fatigue anomalies with the external fuel tanks were cause for more serious concern, and involved manufacturing and

The original external tanks used an elliptical cross section, as shown here. The tanks were constructed using spun fiber impregnated with aluminum, which proved prone to fatigue cracks. A decision was made to switch to more conventional aluminum tanks. Since the elliptical tanks had generated more drag than expected, the new tanks featured a circular cross section. *(Frederick A. Johnsen Collection)*

cost issues as well. Eventually a decision was reached to replace the original construction of spun-fiber-impregnated-with-aluminum with more conventional all-aluminum construction. Since flight testing had also revealed that the original elliptical-cross-section tanks generated more drag than expected, the new aluminum tanks had a more conventional circular cross section. At the same time the capacity of the tanks was increased from 315 to 330 gallons.

During late 1980 the F/A-18 was tested in the Air Force's McKinley Climatic Laboratory at Eglin AFB. The aircraft was exposed to climatic extremes between –65 and +125°F, along with simulated winds up to 100 mph and rains of 20 inches per hour. The normal minor problems (mainly leaks) were found, but all were easily corrected prior to production.

Initial flight testing went well, but some problems were nevertheless encountered. The nosewheel lift-off speed was excessively high and the takeoff roll was too long. These problems were solved by filling in the snag on the leading edge of the stabilators, giving them greater authority at an earlier juncture during the takeoff run. The snag had been added to the leading edge of the stabilator in anticipation of the same flutter problems that had affected the F-15, but these problems never materialized. In addition, a greater upward moment was provided by automatically toeing in the rudders during takeoff, reducing rotation speed, by 30 knots, to 115 knots.

Shortly after the carrier suitability trials, an F/A-18A (BuNo 160777) suffered a main gear failure during touchdown at Pax River. The fault was traced to the axle centering unit, and the fix was to incorporate a larger twin-chamber oleo strut. After the aircraft was repaired with the new design, it was sent to Edwards AFB to conduct crosswind landing tests. A total of 119 landings were completed successfully in crosswinds as high as 30 knots.[56]

The observed 185° per second roll rate at Mach 0.7 and 10,000 feet[57] was significantly below the 280° per second requirement. In fact, the test program experienced roll rates as low as 50° per second at transonic speeds below 5,000 feet, resulting in a major redesign of the wing. Part of the problem was traced to an outer wing panel twist. In other words, the wing flexed too much—as the aileron moved up or down, the wing leading edge would move in the opposite direction, partially negating the rolling moment.

To correct the problem, the ailerons were extended all the way to the wingtips (resulting in 36 percent more area), and differential leading- and trailing-edge flap movement was added to the flight control software. Perhaps most important, the monolithic graphite-epoxy composite construction was replaced by an aluminum and composite sandwich that significantly stiffened the outer wing panel. Along with additional differential stabilator authority, these changes largely eliminated the roll problems. At the same time, the ailerons were programmed to droop 45° during low-speed flight, reducing the approach speed by 10 knots.[58]

A problem was also experienced where the leading-edge flaps locked up under high-g maneuvering. This was traced to the hydraulic drive unit being unable to provide enough torque to drive the leading-edge flaps under high-load conditions. When the flight control computers sensed a discrepancy between the commanded position and the actual leading-edge flap deflection, they commanded the leading-edge flaps to lock in their present position. The fix involved reducing the leading-edge flap chord by eliminating the lead-edge snag, and by increasing the leading edge radius (i.e., making it more rounded).

There were of course other problems. These included angle-of-attack hang-ups, falling leaf mode, throttle friction and linearity, environmental conditioning system (ECS) problems, fuel cooling problems, center of gravity control, high-gain/closed-loop handling qualities, 5.6-hertz limit oscillations, and the loss of the second aircraft due to a catastrophic engine failure. Each of these problems was worked through by the engineers and corrected before production aircraft reached the Fleet.

The F/A-18A, and indeed the follow-on F/A-18C, suffered from insufficient range. In fact, this has been the aircraft's most-often criticized shortcoming, and has never really

The most serious problem afflicting the Hornet was structural cracks in the vertical stabilizer mounts. Beginning in May 1988 a small fence was added to the top of each LEX near the wing leading edge in order to broaden the vortices which were causing the fatigue. Known as *LEX fences*, these metal plates ultimately proved to be an inexpensive solution to the problem. The fences also provided a minor improvement in controllability at high angles of attack. *(Mick Roth)*

been fully corrected despite several engine and airframe modifications. Perhaps the most significant of these changes was an alteration of the boundary layer air discharge (BLAD) slots. The FSD aircraft originally had long BLAD slots cut between the fuselage and the upper surface of the LEXs, much like the YF-17s. These slots had the beneficial effect of stabilizing and helping to control the strong vortex generated by the LEX that extended down each side of the fuselage, and also presenting "clean" air to the vertical stabilizers at high angles of attack, increasing directional stability. Unfortunately, some engineers thought they also generated a great deal of parasitic aerodynamic drag, which adversely affected range and acceleration. Although the drag issue was not fully studied, 80 percent of the length of the slots were filled in beginning with the eighth FSD F-18A, leaving only one small slot on each side whose function was to eject the boundary layer air bled from the engine intake.

The turbulent air caused by the elimination of these slots may have contributed to the fatigue problems experienced by the vertical stabilizers in later years. Aerodynamic loads on the vertical stabilizers from LEX-generated turbulence were particularly severe, and fatigue cracks began to appear at their attachment points. The F-18 fleet was briefly grounded in late 1984 until the aircraft were modified with 4-inch aluminum doublers (referred to as *cleats*) on several of the attachment locations. Changes were made on the production line to ensure the cracks would not appear on new aircraft later in their service lives. All the necessary rework was handled by McDonnell Douglas under warranty. These changes, however, failed to completely alleviate the problem.

Beginning in May 1988 a small fence was added to the top of each LEX near the wing leading edge in order to broaden the vortices which were causing the fatigue. Known as *LEX fences*, these metal plates ultimately proved to be an inexpensive solution to the problem. The fences also provided a minor improvement in controllability at high angles of attack.

The first production F/A-18A (BuNo 161213) made its maiden flight on 12 April 1980.[59] Production was delineated into "Lots," each usually subdivided into three "Blocks." The number of aircraft per block varied, but peaked at 33. A total of 371 production F/A-18As were built before production switched to the F/A-18C in FY86.[60]

F/A-18B Two-Seater

The F/A-18B is the two-seat combat trainer version of the F/A-18A. Two FSD aircraft (BuNos 160781 and 160784) were followed by 39 production F/A-18Bs. Initially designated TF-18As, they were intended primarily as proficiency trainers, but retained full combat capability. To make room for the second seat, internal fuel capacity was reduced by 110 gallons. The rear cockpit was equipped with a throttle and stick, allowing an instructor to fly the aircraft if necessary.

Sled tests at China Lake revealed that the rear ejection seat could not consistently break through the canopy. As a result, all two-seat Hornets were restricted from carrying a back-seat occupant for a period of time. One proposal included adding an explosive cord to the canopy (as on the AV-8B) to shatter it prior to ejection. The idea of sharpening the canopy breakers on top of the seat was discarded as too hazardous, and additional seat propellant was also discounted. Ultimately, a new canopy material that shattered more easily solved this problem.

Initial Service

Following trials at Pax River and follow-on operational test and evaluation by VX-4 and VX-5 at PMTC Point Mugu and NWC China Lake, the Hornet was declared ready for service.

An F/A-18A (forward) and F/A-18B from VFA-125 conduct flight operations aboard the USS *Carl Vinson* (CVN-70). The Hornet reversed a trend toward heavier and heavier fighters, and very few problems were encountered during carrier qualifications. *(U.S. Navy/DVIC DN-SC-84-11525)*

The size difference between the F/A-18 and the F-14 is obvious in this photograph of the USS *Constellation* (CV-64). Of course, the Hornet was designed as the "low" part of the high-low mix, and initial versions were much less capable than the F-14 in low-range air-to-air combat. Even today the Hornet cannot carry the long-range radar and missiles of the F-14, although many argue these are no longer necessary in the post–Cold War world. *(U.S. Navy/DVIC DF-ST-92-07920)*

The Pacific Fleet Readiness Squadron (FRS), VFA-125 "Rough Riders," was commissioned at NAS Lemoore on 13 November 1980, and received its first F/A-18 three months later. The first production aircraft arrived in September 1981, and by the end of the year the squadron had eight Hornets. Eventually the squadron would have up to 60 aircraft assigned to it as it provided conversion training for pilots transitioning from existing Navy and Marine fighter and attack squadrons. Later, the squadron concentrated on training new pilots with no

An F/A-18A waits its turn as steam rises from the other catapult. Carrier operations are the reason American Hornets are only authorized to carry the 330-gallon external fuel tanks, since the Navy believes the 480-gallon tanks represent a clearance problem on the centerline station during cats and traps. *(U.S. Navy/DVIC DN-SC-86-01291)*

A CF-18B prepares to refuel—note the deployed refueling probe. Canada was the first foreign customer to purchase the Hornet, receiving A/B models that were essentially similar to U.S. aircraft. The position of the 20-mm cannon directly in front of the pilot caused some initial concern about gun gases pitting the windscreen during firing, but these proved unfounded. *(Mike Reyno/Skytech Images)*

Fleet experience, and also provided training for most foreign customers. Subsequently, VFA-125 was supplemented by an Atlantic Fleet FRS, VFA-106 "Gladiators" based at NAS Cecil Field, and by a Marine Training Squadron, VMFAT-101 "Sharpshooters," at MCAS El Toro. (VFA is the Navy designation for fighter-attack squadron.)

The initial experience with VFA-125 showed that the concerns about the range of the F/A-18 were somewhat overstated. Nevertheless, the range fell below specification in all

Lieutenant Commander "Warden" Davis from VFA-25 lands an F/A-18A aboard USS *Constellation*. An empty multiple ejector rack (MER) probably indicates he is returning from bombing practice, since it is usually used to carry 25-pound practice bombs. *(U.S. Navy/DVIC DN-ST-84-11928)*

The Royal Australian Air Force also purchased A/B model Hornets. This great study of the underside of an AF-18A (A21-14) from No. 3 Squadron shows the standard air-to-air loadout of four AIM-7s and two AIM-9s. Note that the roundels are both oriented toward the center of the aircraft, not to the direction of travel. Black square just behind and inboard from the fuselage AIM-7 noses are the ALE-39 countermeasure dispensers. *(RAAF/LAC Jason Freeman via Chris Hake)*

Many early Hornets have been used as aggressor aircraft by various squadrons. This F/A-18A (BuNo 162114) was operated by the Navy Strike and Air Warfare Center (NSAWC) at NAS Fallon, Nevada, during September 1998. The upper surfaces were painted in three tones of brown/tan, but the lower surfaces remained overall gray. There were a large number of different paint schemes used by the aggressors, most mimicking Soviet or Eastern Europe camouflage. *(Mark Munzel)*

The Navy Strike Warfare Center "Strike U" [now part of the Navy Strike and Air Warfare Center (NSAWC)] used this F/A-18B (BuNo 161707) to train strike pilots much the same way that Top Gun trained for air-to-air missions. *(Mark Munzel)*

mission configurations. In some instances this shortfall was very small. The requirement for the combat radius in the fighter escort role was 400 nautical miles—the F/A-18A actually achieved 380. The squadron completed carrier qualifications aboard the USS *Constellation* (CV-64) during the first week in October 1982, completing 57 day and 24 night launches and traps, as well as 10 touch-and-go landings.[61]

Interestingly, the Marine Corps put the F/A-18 into operational service before the Navy. The Marines had originally been scheduled to receive 190 F-14As beginning in FY74 for use as their primary fighter. But the Marines had never fully embraced the Tomcat, and when its ground attack capabilities[62] were not developed, the Marines finally opted to purchase 138 additional F-4Js instead. Ultimately the Marines decided to standardize on a force of AV-8 Harriers and F/A-18s. The first Marine units to convert to the Hornet were VMFA-314 "Black Knights," and VMFA-323 "Death Rattlers," both based at El Toro. They received their first F/A-18s in January and March 1983, respectively, and deployed aboard the USS *Coral Sea* (CV-43) for its 1985/86 Mediterranean cruise.

The reports back from the cruise showed just how good the F/A-18 was. From the beginning, the Navy had wanted an aircraft that was more reliable and easier to maintain than anything else in the Fleet. Not just a little better—a lot better. The initial operations with the Hornet showed they had achieved this. The radar worked all the time, every time. This was a major change from the F-4J when many missions were launched without a working radar, and most missions were recovered without one. The weapons carriage/separation system worked every time, and seldom did an F/A-18 bring back a "hung" store. Turnarounds between missions usually involved rearming and refueling since nothing needed to be fixed because nothing was broken. On most pre-Hornet cruises, all the operational aircraft were kept on the flight deck, along with many broken ones. The hangar deck was usually filled with aircraft being cannibalized to keep the few operational ones working. On the initial *Coral Sea* cruise, the hangar deck had operational aircraft stored on it because there was no more room on the flight deck. There were very few broken aircraft. The Fleet was impressed.

The upper surfaces of an Australian AF-18A from No. 2 Operational Conversion Unit (OCU) Squadron. Note that the Royal Australian Air Force (RAAF) does not use national insignia on the upper surfaces. The location of the LEX fences is clearly visible. *(RAAF/LAC Jason Freeman via Chris Hake)*

Navy squadrons VA-113 "Stingers" and VA-25 "Fist of the Fleet" at NAS Lemoore converted from A-7Es to F/A-18As in March and July 1983, respectively. The squadrons were redesignated VFA-113 and VFA-25, reflecting their new multirole capabilities. The first operational cruise by Navy Hornet squadrons took place in February to August 1985 when VFA-113 and VFA-25 deployed aboard the USS *Constellation* to the Western Pacific and Indian Oceans. This cruise confirmed the Hornet as an extremely reliable aircraft requiring less maintenance than the F-14A or the A-6E. Mission capable rates during the cruise averaged 89 percent.

The U.S. Naval Test Pilot School operates four two-seat Hornets including this F/A-18B (BuNo 161360). The aircraft is overall gloss white with a red strip. *(Bill Kistler)*

Libya: The Hornet's First Sting

During 1986, Libya's Colonel Khaddafi claimed the Gulf of Sidra as Libyan territorial waters, declaring a "line of death" across the entrance to the gulf beyond which ships of other nations would not be allowed to enter. In response, President Ronald Reagan ordered the Sixth Fleet to begin "freedom of navigation" maneuvers in the Gulf of Sidra to demonstrate American resolve to operate freely in what it believed to be international waters. The *Coral Sea* (CV-43) was deployed to the region, supplemented by the USS *America* (CV-66) and *Saratoga* (CV-60) for periods of the operation.

During the mid-1980s, it was normal for U.S. aircraft carriers to put to sea with two fighter squadrons (F-4 or F-14) and two attack squadrons (A-6 or A-7). But the *Coral Sea* was too small to effectively operate the F-14, and the Navy decided to prototype a force using four F/A-18 squadrons and a single A-6E squadron. Although the Hornet was joining the fleet in ever-increasing numbers, the Atlantic Fleet did not have four operational Hornet squadrons. The solution was to call on the Marine Corps. Carrier Air Wing Thirteen (CVW-13), led by Commander Byron Duff, therefore comprised Navy squadrons VFA-131 "Wildcats" and VFA-132 "Privateers" along with Marine squadrons VMFA-314 and VMFA-323, plus one Navy A-6 attack squadron. F/A-18s from the *Coral Sea* flew combat air patrols, protecting the carrier group from Libyan MiG-23s, MiG-25s, Su-22s, and Mirage F.1Qs sent to observe the Fleet. The Hornets also flew suppression of enemy air defenses (SEAD), surface combat air patrol (SUCAP), and fleet air defense missions.

After a terrorist bombing in a Berlin disco was traced to Khaddafi, Reagan decided to take action. Operation Prairie Fire ran from 24 March through 14 April 1986, and involved several air strikes against Libyan shore installations. During this action, F/A-18As attacked an SA-5 missile site at Sirte which had been tracking U.S. aircraft. This marked the combat debut for the Hornet, and was also the first operational use of the AGM-88A high-speed antiradiation missile (HARM). The Hornets attacked the SAM site in deteriorating weather from very low level, and all of the F/A-18As returned without mishap.

Most of the A/B models that are still flying with the U.S. services are in the Reserve squadrons, such as this F/A-18A (BuNo 162453) with VFMA-112 at JRB (Joint Reserve Base) Fort Worth (the former Carswell AFB). These aircraft will likely be replaced by early C/D models as the E/F model enters active service. *(Mark Munzel)*

An F/A-18A shows an AGM-62 Walleye under the wing during June 1993. The two small slots on top of the LEXs near the dorsal spine are all that are left of the original BLAD slots that originally extended all the way to the cockpit. Filling them in provided a significant decrease in drag without severely affecting high angle-of-attack handling. *(Mick Roth Collection)*

Another CF-18B (188910), this one showing how far the nose landing gear can rotate in the "low" setting. Some staining can be seen around the gun gas exhaust ports on the upper nose. *(Mark Munzel)*

The first AF-18B (A21-101) participated in weapons testing. Note the cameras under the rear fuselage and wingtips. *(Tony Landis Collection)*

No sooner had Operation Prairie Fire ended than Operation Eldorado Canyon began. On 15 April 1986, a combined Air Force/Navy attack took place against Libyan targets. Aircraft, including Hornets, from the *Coral Sea* were dedicated to attacking targets around Bengazi, while aircraft from the other carrier on station and Air Force F-111F interdiction aircraft from bases in Great Britain attacked Tarabulus (Tripoli).

Near midnight on 14 April, as strike forces headed south, eight F/A-18s armed with AGM-88As were launched from the *Coral Sea*, while another eight Hornets were launched for combat air patrol (CAP) and rescue combat air patrol (RESCAP) duties. The Hornets stayed off the coast as the A-6 strike force went "feet dry" to press its attack. Several EA-6Bs were well positioned to negate any attempt to utilize the Libyan fire control radars, so although several SA-2s were launched, they appeared to be unguided. The A-6 strike force encountered heavy SAM activity over downtown Benghazi, with threats from SA-2, SA-3, SA-6, and SA-8 SAMs and ZSU-23 AAA, but none of the aircraft were hit and the F/A-18s off the coast fired 16 AGM-88As in return. Admiral Frank Kelso, the Sixth Fleet commander, later stated that ". . . nobody has ever flown a mission in a more dense SAM environment." The SA-5 site at Sirte, which had been attacked during the raid in late March, activated its radar as strike aircraft were pulling off their targets, but no SAMs were fired and the Hornets did not engage.

Blue Angels

In February 1986 the F/A-18A was selected to replace the A-4Fs used by the Blue Angels flight demonstration team. The Hornet's public debut with the team was at MCAS Yuma, Arizona, on 25 April 1987. The team initially received eight single-seat F/A-18As (BuNos 161520, 161521, 161523-161527, and 161588), and one two-seat F/A-18B (BuNo 161355).

The Blue Angels' aircraft are, of course, always absolutely cosmetically perfect. The LEX makes a convenient place for pilots to stand as they ingress the aircraft. Note the unusually casual pose of the ground crewman near the access ladder, while the one partially hidden on the other side of the aircraft is at rigid attention. *(Bill Kistler)*

Although the Blue Angel's aircraft are all early examples of the Hornet and have been stripped of most of their operational equipment, they are still capable of operating from aircraft carriers, and do periodically just to prove it (and to keep the pilots proficient in the technique). *(Mark Munzel)*

The two solo aircraft (Nos. 5 and 6) entertain the crowd while the diamond is getting ready for its next maneuver. Here No. 6 streaks at high speed and low level past the flight line. *(U.S. Navy/DVIC DN-ST-89-07207)*

The aircraft were early F/A-18s, which were no longer considered capable of routine carrier operation, and were equipped with new flight control system software optimized for aerobatics. The gun and radar were removed, new seat harnesses were installed to help the pilot handle the weightlessness caused by some maneuvers, and civilian ILS and navigation equipment was fitted. A smoke generation system was installed for use during aerial displays.[63]

Yes Virginia, the Blue Angels do switch numbers between aircraft. Or occasionally, they end up with two carrying the same number. Originally only a single F/A-18B was assigned to the squadron, mainly to give VIP rides before shows. For the past several years the squadron has had a pair of two-seaters assigned. This photo was taken in October 1997 when they performed at Biggs AAF near El Paso. *(Mark Munzel)*

Over the years the squadron has operated various Hornets, all early A/B-models. In late 1999 the Blue Angels operated nine F/A-18As (BuNos 161948, 161955, 161959, 161962, 161963, 161975, 161983, and 161984) and two F/A-18Bs (BuNos 161932 and 161943). All of these aircraft are from Lot VI. Of the nine single-seaters, six perform in the show and one is usually kept as a spare, while the other two are undergoing maintenance. It should be noted that Blue Angels tail numbers are often switched between aircraft, depending upon their maintenance status.[64]

When Boeing and McDonnell Douglas merged in 1997 it posed a unique challenge for the Blue Angels. The F/A-18s operated by the squadron were emblazoned with "McDonnell Douglas" on the fuselage, and the pilots used the lettering to line up their aircraft during formation flying. Keeping the leading edge of the wing aligned with the correct letter on the opposite aircraft allowed team members to keep the proper distance, elevation, and separation during formation maneuvers. When it came time to change to "Boeing," the Blue Angels needed to determine exactly what dimensions would work for the new lettering and exactly where the decal would be placed on the aircraft. Boeing provided the new decals to the government at no cost.[65]

On 28 October 1999, a Blue Angels F/A-18B (BuNo 161932) crashed while preparing for an air show at Moody AFB, Georgia. The aircraft crashed while performing a "circle and arrival" maneuver, whereby the team flies over an airfield to familiarize itself with the local area. Lieutenant Commander Kieron O'Conner and Lieutenant Kevin Colling were killed in the crash. It was the team's first fatalities in 14 years, and the first crash of any kind in 9 years.

The show at Moody and two shows at NAS Jacksonville, Florida, scheduled for the following weekend were canceled. The team returned to the air for the season's last two shows at NAS Pensacola, Florida, before a crowd of 55,000. The shows were performed with five instead of the usual six aircraft as a tribute to O'Conner and Colling. A missing man formation was flown at the beginning of each show. Rear Admiral Michael Bucchi, the commander of the Naval Air Training Command to which the Blue Angels belong, flew in the back seat of one aircraft during the first show.

Hornets for NASA

Beginning in late 1984, the NASA Dryden Flight Research Center[66] (DFRC) took delivery of several F/A-18A/Bs for chase and proficiency flying, replacing F-104 Starfighters. NASA's experience with the F/A-18 did not start out particularly well. The aircraft were all FSD and early production examples, and were not in terribly good shape when they arrived at Dryden. Since many of the aircraft were essentially prototypes, few had been built to the same configuration, and many of the subassemblies and systems were not production units and could not be requisitioned through the normal supply channels. It took the engineers and technicians at Dryden a considerable amount of effort to restore the initial aircraft to flyable condition. The Navy was happy to loan the FSD aircraft to NASA "because their unique configuration makes them unsuitable for fleet use . . . they were becoming difficult and costly to support," and they "have about 20 percent parts commonality with subsequent production aircraft."

Two of the aircraft that were delivered were used as sources of spare parts, and the airframes were later returned to the Navy without ever having been assigned NASA numbers. NASA 842 was retired from flight status and put on display in front of the Lancaster Jethawks sports stadium painted in NASA markings. In addition, one aircraft (BuNo 161251) is used as the "Iron Bird" in-loop simulator in the Research Aircraft Integration Facility (RAIF) and was not assigned a NASA number since it is not flyable. The other NASA aircraft are shown in the following table:[67]

Tail No.	Type	BuNo	Acquired	Notes
None	F/A-18A	160775	3 Jan 85	Used for spares. Returned to Navy.
None	F/A-18A	160785	22 Oct 84	Used for spares. Returned to Navy.
None	F/A-18A	161251	13 Dec 85	Used as "Iron Bird" in RAIF.
NASA 840	F/A-18A	160780	22 Oct 84	HARV. Currently in storage at DFRC.
NASA 841	F/A-18A	161216	1 Oct 85	Retired to the range at China Lake.
NASA 842	F/A-18A	161214	24 Aug 87	On display at Lancaster Sports Stadium.
NASA 843	F/A-18A	161250		Returned to Navy.
NASA 843	F/A-18A	161519	1 Dec 91	Second aircraft called NASA 843.
NASA 844	F/A-18A	161213	1985	Crashed on 7 Oct 1988.
NASA 845	F/A-18B	160781	21 Jul 86	Systems Research Aircraft (SRA).
NASA 846	F/A-18B	161355	1 Mar 91	Chase/support.
NASA 847	F/A-18A	161520	21 Sep 89	In storage at DFRC.
NASA 848	F/A-18A	161949	28 Dec 89	Returned to Navy on 6 Nov 90.
NASA 850	F/A-18A	161703		Chase/support.
NASA 851	F/A-18A	161705		Chase/support.
NASA 852	F/A-18B	161217		Chase/support.
NASA 853	F/A-18A	161744	4 Mar 99	Advanced Aeroelastic Wing (AAW).

When the first F/A-18A (BuNo 160775) was delivered to NASA, it still carried the modified reconnaissance nose it had been testing at Pax River. The aircraft was used for spare parts to get the other NASA F/A-18s flying. *(Tony Landis Collection)*

NASA donated one of its F/A-18As (BuNo 161214) to the Lancaster sports stadium as a display. Here the aircraft is being lifted into place. *(NASA photo by Tony Landis)*

The first mission assigned to the NASA Hornets was to provide proficiency training for pilots of the Grumman X-29 forward-swept-wing research aircraft. The Hornets also frequently flew chase for the X-29 since their medium-speed handling was similar. F/A-18s were also used as chase aircraft for the laminar flow experiments conducted with a borrowed F-14. One of the Hornets was dedicated to the High Angle-of-Attack Research Vehicle (HARV) research program, while a second was modified to serve as a Systems Research Aircraft (SRA) for a variety of experiments. The remaining aircraft perform chase and proficiency training duties.

On 7 October 1988 NASA test pilot Stephen Ishmael found himself in an uncontrollable spin in NASA 844, caused by an asymmetrical leading-edge flap deployment. Ishmael elected to eject from the aircraft, and was somewhat bruised and battered in the process. Fortunately, he was able to return to flight status soon afterward.

HARV

Dryden used NASA 840 as its High Angle-of-Attack (Alpha) Research Vehicle in a three-phased flight research program that lasted from April 1987 until September 1996. The aircraft completed 385 research flights and demonstrated controlled flight at angles of attack between 65° and 70° using thrust vectoring vanes, a research flight control system, and (eventually) forebody strakes.[68]

Angle of attack (alpha) describes the angle of an aircraft's fuselage and wings relative to its actual flight path. During maneuvers, pilots often fly at extreme angles of attack with the nose pitched up while the aircraft continues in its original direction. This can lead to conditions in which the airflow becomes separated over large regions of the lifting surfaces, resulting in insufficient lift to maintain control of the aircraft and a corresponding increase in drag, leading to a stall.

Chosen partially because it was available, but also because it exhibited excellent high-angle-of-attack capabilities, this F/A-18A (BuNo 160780) was used as the High Angle-of-Attack Research Vehicle (HARV). The aircraft made 385 flights during the HARV test program before structural cracks were discovered that proved too expensive to repair. The aircraft will be put on display at Dryden, and its place in the research flight taken by NASA 853 (BuNo 161744). *(NASA photo by Tony Landis)*

During 1985, NASA began a test program to explore and understand aircraft flight at very high angles of attack. NASA-Langley managed this High Angle-of-Attack Technology Program (HATP) in partnership with the Ames, Dryden, and Lewis research centers. The HATP produced technical data from actual flight at very high angles of attack to validate computational fluid dynamics (CFD) and wind tunnel research. Successful validation of these data has given engineers a better understanding of aerodynamics, effectiveness of flight controls, and airflow phenomena at high angles of attack. Motivated by the tactical advantages of enhanced high angle-of-attack agility and maneuverability, they have used this understanding to design features providing better control and maneuverability in future high performance aircraft.

The F/A-18A had no angle-of-attack restrictions at normal center-of-gravity positions, making it a logical choice as a research vehicle. The F/A-18A (BuNo 160780) selected by NASA had been used during the spin recovery program at Pax River and was equipped with an emergency recovery system. The aircraft had only 400 flight hours on it when NASA acquired it but, unfortunately, had been heavily "cannibalized" (used as a source of spare parts) by the Navy, which never expected the aircraft would fly again. It arrived at Dryden in pieces aboard a truck in October 1984, missing 400 parts and having very little documentation on its existing wiring system. Dryden mechanics and technicians had to find substitute parts, cut out all the existing wiring, then assemble the aircraft and rewire it.

NASA 840 required 18 months of refurbishment prior to beginning HARV flights in 1987. Because of this it was named "The Silk Purse," as in "making a silk purse out of a sow's ear." The name was painted in red letters below the canopy. When the aircraft was painted in the HARV scheme, it was changed to "Silk Purse" in gold script.[69]

The aircraft was equipped with camera pods on the wingtips in lieu of the Sidewinder rails to photograph streams of white smoke emitted from the forward fuselage and LEX to highlight airflow patterns. To make the smoke trails stand out better, the upper surfaces of the aircraft were painted black. In order to provide details about on-surface flow patterns, a special red-dyed glycol solution could be emitted from dozens of tiny holes in the aircraft's nose and filmed as it streamed over the surface of the fuselage. Short pieces of yarn (tufts) were also glued to various parts of the aircraft to provide additional visual clues.

HARV Phase One The first phase of "high alpha" flights began in April 1987 and lasted through 1989. It consisted of 101 research flights at angles of attack as high as 55°. The purpose of this phase was to obtain experience with aerodynamic measurements at high angles of attack and to develop the necessary flight research techniques. During this phase, there were no external modifications to the aircraft, although it was equipped with extensive instrumentation. NASA research pilot Einar Enevoldson made the first functional check flight on 2 April 1987 and three succeeding flights before turning the piloting duties over to Bill Dana and Ed Schneider.

HARV Phase Two The second phase examined the benefits of using vectored thrust to achieve greater maneuverability and control at high angles of attack while continuing the correlation of flight data with wind-tunnel and CFD data begun in Phase One. This phase featured major hardware and software modifications including a multiaxis thrust-vectoring system and a research flight control system. Three paddlelike vanes made of Inconel-625[70] were mounted around each engine's exhaust to provide pitch (up/down) and yaw (right/left) control when the aerodynamic surfaces were either unusable or less effective than desired. The vanes used modified aileron actuators.

The engines had the divergent portion of the exhaust nozzles removed to shorten the distance the vanes had to be cantilevered by about 2 feet. The subsonic performance of the engines, including afterburning, was largely unaffected by the modifications, but supersonic flight was no longer possible. The thrust vectoring system added 2,200 pounds, while the spin recovery parachute, emergency power system, and other equipment added a further 1,919 pounds. In general, the thrust vectoring system was positioned to avoid interference with the aerodynamic control surfaces. However, the inside trailing edges of the stabilators were cut away slightly to provide clearance for the lower outboard vane actuator housing.

Research flights using the thrust vectoring system began in July 1991 and concluded in January 1993. The system resulted in significantly increased maneuverability at moderate angles of attack and some degree of control at angles of attack up to 70°. It also allowed researchers to collect a greater amount of data by remaining at high angles of attack longer than they could have done without it. The pilot used standard cockpit controls, and no special pilot action was required after the system was engaged in flight.

Between January 1993 and January 1994, the aircraft was modified with additional instrumentation, including a sophisticated pressure measurement system between the inlet entrance and the engine face. This information provided unprecedented understanding of what happens to engine airflow under extreme maneuver conditions. Flights resumed in late January 1994 and continued through June 1994, with Ed Schneider and Jim Smolka as the Dryden research pilots, joined for short periods by Navy guest pilots. There were a total of 193 flights during Phase Two.

HARV Phase Three During March 1995 Dryden installed movable strakes on both sides of the aircraft's nose to provide yaw control at high angles of attack where conventional rudders became ineffective. These strakes, 4 feet long and 6 inches wide, were hinged on one side and mounted to the forward sides of the fuselage. At low angles of attack, they were folded flush against the aircraft skin. At higher angles of attack, they were extended to interact with the strong vortices generated along the nose and thereby produce large side forces for control. Wind tunnel tests indicated strakes could be as effective at high angles of attack as rudders are at lower angles. Similar strakes had been tested on an F-14 at Dryden 10 years earlier.

Flights with the strakes began in July 1995, and the strakes enabled pilots to employ three separate flight modes. One used thrust vectoring alone. Another used thrust vectoring for longitudinal (pitch) control and a blend of thrust vectoring and strakes for lateral (side-to-side) control. The third mode used thrust vectoring solely for longitudinal control, with strakes alone for controlling lateral motion. The strake project concluded in September 1996 on the 109th flight of Phase Three. It yielded a great deal of information about the operation of nose strakes, which were effective in providing control above 35° angle of attack. Besides Schneider and Smolka, research pilots included Mark Stucky from Dryden, Phil Brown from Langley, and a number of guest pilots from the Navy and Marine Corps, Canadian Forces, Royal Air Force, McDonnell Douglas, and CalSpan. There had been a total of 385 flights during the HARV program.

It had been expected that NASA 840 would be used for follow-on research into an active aeroelastic wing (AAW), but in June 1999 while the aircraft was being modified to incorporate the new wing, severe cracking of the vertical stabilizer mounts was discovered. A careful examination revealed that the structural problems were too extensive to repair, and the aircraft was retired from flying status. The aircraft will be repainted as HARV and placed on display at Dryden. Instead, NASA 853 (BuNo 161744) will be modified by using systems from HARV to continue its research.

Systems Research Aircraft

Dryden uses an F/A-18B (BuNo 160781) as its Systems Research Aircraft (SRA).[71] The SRA is used to identify and flight-test high-leverage technologies beneficial to subsonic, supersonic, hypersonic, or space applications. Key technologies investigated aboard the F-18 SRA include advanced power-by-wire concepts, electric-powered actuators and mechanical systems, fly-by-light (fiber-optic) systems, and advanced computer architectures. In addition, the program is developing advanced flight-test techniques that will be used on future aircraft.[72]

Improved Hornets: The F/A-18C/D

The more capable F/A-18C quickly began to replace the A/B in front-line squadrons. Partly in anticipation of the F/A-18C/D's impending Fleet arrival, Naval Air Reserve squadrons began receiving F/A-18As in September 1985. The first was VFA-303 "Golden Hawks," based at NAS Lemoore. But even as 1999 drew to a close, the Marines still had two operation squadrons using the A/B, and there is one more A-model carrier deployment planned for late 2000.

The differences between the F/A-18A and C were almost entirely internal. The F/A-18C featured a revised Martin-Baker NACES ejection seat, improved XN-6 mission computers, an upgraded stores management set, an upgraded armament bus (MIL-STD-1553B and -1760 compatible), and a flight incident recorder and monitoring set (FIRAMS). The FIRAMS incorporates an integrated engine/fuel indicator, performing continuous monitoring and recording of aircraft maintenance and flight incidents data, and a processor that provides overall fuel system management including center-of-gravity maintenance. External differences were limited to various blisters to cover the antennas for the AN/ALQ-165 advanced self-protection jammer (ASPJ).

In reality, the C/D bears little resemblance to the A/B from a mission perspective. The aircraft systems were designed to be upgradable, and most of the systems are digital, communicating over standard data buses. In addition, a majority of the systems are software-driven, so they are easy to update.

All Navy aircraft use versions of the Control Data International[73] AYK-14 standard mission computer. This has greatly simplified programming requirements, maintenance, and logistics for naval aircraft. However, this is also overstating the case. Although all of the computers are designated AYK-14, in reality the processors themselves have changed greatly over the last 20 years—other than the basic interfaces, very little is common between the original units and the latest versions. The XN-6 mission computer originally used in the F/A-18C/D offered 3 times the processing speed and double the memory of the XN-5 used in the A/B. Beginning with FY91 production, the XN-6 was replaced by the even more capable XN-8, and current production aircraft use XN-10s.[74]

While the major differences between the A/B and C/D models are internal, there are a few exterior clues. New ECM antennas were added on the trailing edge of the vertical stabilizers, on the dorsal spine just aft of the cockpit, and on the forward fuselage just under the tip of the LEX. Later ECM modifications would add more antennas. *(Mark Munzel)*

Barely visible above the formation light strip on the nose is a blister that covers the ALQ-165 forward upper high-band transmit antenna. When the Navy decided to use ALQ-165 on F/A-18C/Ds operating in high-threat areas, the existing ALQ-126 systems were removed. This has created an interesting problem, since when the squadrons rotate out of the threat area, the ALQ-165 is removed because there are not enough systems in the inventory to equip all aircraft. However, all of the antenna blisters remain on the aircraft. This F/A-18C (BuNo 164197) is from VFA-105. *(Mick Roth Collection)*

Flight deck crewmen signal for an F/A-18C from VFA-86 to move into position on the No. 1 catapult aboard USS *America* (CV-66) during Operation Deny Flight. The way the ailerons droop to form what are effectively full-span trailing-edge flaps shows up well here. *(U.S. Navy/DVIC DF-ST-95-00560)*

The first production F/A-18C (BuNo 163427) made its maiden flight on 3 September 1987 with Glen Larson at the controls, and on 23 September 1987 the aircraft was flown by Lieutenant Commander John Bell to the Naval Weapons Center at China Lake. Testing went quickly, and since no major changes were required on the production line, the F/A-18C quietly entered full-scale production in mid-1987. It would take another 18 months before Fleet squadrons would transition to the F/A-18C, the first being VFA-25 "Fist of the Fleet" and VFA-113 "Stingers." Production F/A-18Cs were initially powered by the same F404-GE-400 that powered the F/A-18A. In the end, 137 basic F/A-18Cs were procured for the U.S. Navy and Marine Corps.

Not surprisingly, during operational test and evaluation the F/A-18C/D showed the same range deficiencies as previous versions of the F/A-18. This resulted in a recommendation from the Navy operational test and evaluation (OT&E) organization to either

The precision necessary to land an aircraft on a moving aircraft carrier is shown here—a few feet too low and the Hornet would crash into the ramp (rear edge) of the flight deck instead of landing upon it. This F/A-18C from VFMA-312 is landing aboard USS *Theodore Roosevelt* (CVN-71). *(U.S. Navy/DVIC DF-ST-97-01190)*

This F/A-18C (BuNo 164669) from VFA-15 is carrying a Walleye under the wing while participating in exercises at NAS Fallon in November 1994. *(Michael Grove via the Mick Roth Collection)*

increase the F/A-18 fuel capacity or provide additional embarked air wing aerial refueling assets. Additionally, the APG-65 failed to meet several classified requirements, although these were reportedly corrected in the APG-73 upgrade.[75]

Many running changes were introduced during F/A-18C/D production, with many oriented toward further improving the mission-capable rate and reducing maintenance needs. For instance, beginning with Lot XIII an on-board oxygen generating system

A line of Hornets from VFA-15, VFA-87, and VMFA-312 parked on the flight deck of the USS *Theodore Roosevelt* during a 1997 exercise. All of the F/A-18s have been modified with the ALQ-165 blisters. *(U.S. Navy/DVIC DF-ST-97-01190)*

A pair of F/A-18Cs from VFA-86 maneuver over the Townsend Target Range. The nearest aircraft is the CAG, an outdated acronym for Commander Air Group, as shown by the use of the "x00" modex. The CAG used to be the most colorful aircraft in the squadron, wearing special markings at the discretion of the commanding officer. These days the Navy officially discourages flamboyant CAG paint schemes, although some squadrons still manage to sneak a little color onto them. *(U.S. Navy/DVIC DN-SC-89-05116)*

(OBOGS) replaced the previous liquid oxygen (LOX) converter system.[76] The Hornet has already established itself as the most reliable and lowest-maintenance fighter in the Fleet, largely because it is the most modern. The Hornet flies an average of 1.8 hours between failures, compared to only 0.5 hour for the F-14. It requires 17.3 maintenance man-hours (MMH) per flight hour, compared to 46.5 for the F-14 and 44.4 for the recently retired A-6E. And Hornets fly a lot. The 1,000,000th flight hour was reached on 10 April 1990; the 2,000,000th on 17 September 1993.[77] The 3,000,000th flight hour came 3 years later, on 23

An example of a more colorful CAG is this F/A-18C (164212) from VFA-131 photographed at NAS Oceana on 29 October 1996. All of the markings are red, white, and blue. The other squadron aircraft carried the same tail markings in subdued shades of gray. *(Craig Kaston via the Mick Roth Collection)*

One of the missions assigned to the F/A-18C is the suppression of enemy air defenses (SEAD) using the AGM-88 high-speed antiradiation missile (HARM), shown on this VFA-81 aircraft. The HARM is the third major antiradiation missile to see Navy service and has proven to be a very capable missile, although it is constantly being improved upon. *(Michael Grove via the Mick Roth Collection)*

December 1996.[78] During the first 1.5 million flight hours, the U.S. lost 67 Hornets[79] in accidents, compared to 124 F-14s and 102 A-6s.[80] However, demonstrating the overall safety record of the F/A-18, on 30 July 1999 VFA-125 surpassed 130,000 flight hours[81] without a major accident.[82]

Night Attack Hornets

Beginning with the FY88 procurement, all F/A-18C/Ds were equipped to full night attack standard and carry the Night Attack Hornet moniker. New equipment includes GEC-Marconi AN/AVS-9 (MXV-810) Cat Eyes night vision goggles (NVG) and gold-tinted

Red-shirted ordnancemen load a Mk 83 bomb onto the inboard pylon of an F/A-18C during Operation Desert Storm. An AIM-7 Sparrow has already been loaded onto the air intake station. Loading bombs is still largely a manual process, especially aboard carriers. *(U.S. Navy/DVIC DN-ST-92-07894)*

An F/A-18C from VFA-137 is pushed into its parking place aboard USS *Constellation*. Note the spotter kneeling on the edge of the flight deck to help the deck crew guide the aircraft into its proper position (without pushing it over the side). Space aboard carriers is very tight, and most aircraft are parked with significant portions of the rear fuselages hanging over the sides. *(Mark Munzel)*

canopies designed to reflect radar and laser energy away from the interior of the cockpit. The Night Attack Hornet uses two Kaiser 5 × 5-inch color multifunction displays (MFD), replacing the monochrome MFD used in previous F/A-18s. The aircraft are also equipped with a Smiths Srs 2100 color digital moving map display, which provides geographical data from an on-board CD-ROM. Also integral to the Night Attack capability is the Raytheon AN/AAR-50 navigation FLIR (NAVFLIR) carried on the right fuselage station. NAVFLIR imagery can be projected onto the heads-up display (HUD) in a 1:1 scale, giving pilots an

One of a variety of paint schemes that have been worn by the VX-9 CAG aircraft. The left side says NAVY, while the right side says MARINES, indicating the joint-service nature of the operational test and evaluation squadron that was created by combining the assets of VX-4 and VX-5. Barely visible under the wing is a GBU-32 JDAM. *(Mick Roth)*

VX-9 conducted tests of the joint direct attack munition (JDAM) during October 1997. Essentially JDAM is a conventional Mk 80 series "dumb" bomb with a Global Positioning System (GPS) receiver and a set of movable tail fins. The aircraft downloads the GPS coordinates of the desired target over the MIL-STD-1760 interface and releases the bomb. The bomb corrects its flight path on the way to the target. This is a relatively inexpensive way to get a "semiprecision" weapon that is far more accurate than a conventional bomb, but somewhat less so than a laser-guided bomb. *(Mick Roth)*

outside view of the world at night. Upgraded NVIS (night vision intensification system)–compliant cockpit lighting makes the instrumentation readable through the green-hued vision of the NVG (although cockpit instruments are often read by glancing beneath the goggles themselves).

Operational evaluations of the original AAS-38 FLIR pod had indicated the sensor was deficient in terms of image magnification and resolution. The evaluations also pointed out the handicap of not having an on-board laser designator. A study of possible upgrades to the AAS-38 showed there was not an economical way to enhance its performance, so the

This F/A-18C (163495) from VFA-83 is carrying an AGM-84E SLAM and its associated data link pod. The SLAM is a derivative of the Harpoon antiship missile, and has provided the Navy with an effective standoff weapon with minimal development costs. *(Vance Vasquez via the Mick Roth Collection)*

A high-speed camera captures a grainy image of an AGM-84E SLAM striking its target. This was the first launch of a SLAM from an F/A-18 guided by its Walleye video data link. This missile was unarmed, but normally the warhead is set to explode after the missile penetrates a short distance into the target, ensuring maximum destruction. *(U.S. Navy/DVIC DN-SC-89-09948)*

pod was released for Fleet use with some restrictions. However, technology advances quickly. Beginning in January 1993, the AAS-38A pod gained a laser target designator/ranger (LTD/R) subsystem, allowing Hornets to deliver precision laser-guided weapons autonomously of an external laser source. A further improved AAS-38B adds a new Texas Instruments signal processor and a laser spot tracker, and also introduces a true air-to-air infrared search capability.

The remaining sensor performance issues with the existing FLIR are being addressed with the follow-on advanced targeting forward-looking infrared (ATFLIR), intended for both the F/A-18C/D and the E/F.[83] The ATFLIR upgrade incorporates "GEN III" midwave infrared (MWIR) staring focal plane technologies intended to maximize air-to-ground targeting performance. Testing began in FY99, with a Fleet introduction projected for FY02. The total cost of the ATFLIR upgrade program is estimated at $1,900 million including research and development, with each of the 547 units costing $1.5 million.[84]

The first Night Attack Hornet made its maiden flight on 6 May 1988, and production deliveries began on 1 November 1989 with the first Lot XII F/A-18C. Deliveries to Navy squadrons began on 18 November 1989, with aircraft going to VFA-146 "Blue Diamonds" at NAS Lemoore. The first Marine Night Attack Hornet was delivered to VMFA-312 "Checkerboards" on 8 August 1991. A total of 330 Night Attack F/A-18Cs were delivered before production switched to the F/A-18E/F.[85]

Beginning in January 1991 the F/A-18C switched to the F404-GE-402 enhanced performance engine (EPE) rated at 17,600 lbf in afterburning, compared to only 15,800 lbf for the earlier -400 engines. The F404-GE-402 was developed in response to a Swiss requirement for additional power, and after reviewing the performance increase afforded by the new

The CAG from VFA-151 shows the new style UHF/VHF antenna being installed on F/A-18s late in their careers. The new antenna (on top of the fuselage) is slightly taller and narrower than the original blade antennas (barely visible under the fuselage). The antennas are functionally identical and are replaced only as needed. Similar antennas are also being used on the F-14, F-15, and F-16. *(Michael Grove via the Mick Roth Collection)*

powerplant, the United States decided to introduce it beginning at Block 36. The performance improvement with the new engine is greater than the numbers would indicate. Because more thrust is available in portions of the flight regime where it is needed most, the EPE has 18 percent more specific excess power at Mach 0.9 and 10,000 feet, and transonic acceleration (from Mach 0.8 to Mach 1.6 at 35,000 feet) is 27 percent better. A typical runway-launched intercept to Mach 1.4 at 50,000 feet takes 31 percent less time with the EPE, and an F/A-18C can be airborne with only a 1,700-foot roll when carrying a standard air-to-air load (two AIM-7s, two AIM-9s, and 500 rounds of 20-mm) at a gross weight of 37,000 pounds.[86]

In 1989 a radar upgrade (RUG) was initiated as a codevelopment program with Canada to improve the electronic counter-countermeasures (ECCM) performance and growth potential of the APG-65 radar. The APG-73 uses the same antenna and transmitter mated with entirely new electronics. The receiver/exciter unit provides much faster analog-to-digital conversions, and the new radar data processor replaces two separate units (the signal processor and

The commander of VFA-136 wants the world to know his squadron is the Knighthawks, as the word emblazoned across his fuselage and fuel tanks indicates. Ground crew are refueling the aircraft using the single-point receptacle on the right side of the fuselage. *(Michael Grove via the Mick Roth Collection)*

the data processor) in the old radar and has over 10 times the processing power. A new power supply that uses solid-state techniques provides a cleaner and more reliable power source.

An original $65.7 million contract was awarded to Hughes in May 1990 for initial development, and an additional $267 million was awarded in June 1991 for continued development and the first 12 production units. The prototype APG-73-equipped F/A-18 made its maiden flight on 15 April 1992, and this radar became standard in all new production C/Ds, including those ordered by Finland, Malaysia, and Switzerland. The first APG-73-equipped Hornets were delivered to VFA-146 "Blue Diamonds" and VFA-147 "Argonauts" at NAS Lemoore on 24 and 25 May 1994, respectively.[87]

The RUG Phase II adds a high-resolution synthetic aperture radar (SAR) mode, provides high-resolution ground mapping for reconnaissance missions, and adds autonomous targeting for the AGM-154 joint stand-off weapon (JSOW) and GBU-32 joint direct attack munition (JDAM). This is also the radar used in the E/F model. The total cost of modifying 547 radar sets through Phase II is estimated at $3,500 million, including associated research and development. The unit cost of the upgrade is approximately $2.1 million.

RUG Phase III includes an advanced electronically scanned antenna (AESA) radar that features greatly improved reliability and maintainability, improved ECCM performance, near simultaneous multimission capability, and enhanced signature characteristics. Since the AESA is physically fixed in one position, the entire drive system can be eliminated, making the unit much lighter, and the antenna will be shaped to improve the radar cross section of the aircraft.[88] Northrop Grumman (ex-Westinghouse) and Raytheon competed for the $1,000 million contract to design and manufacture the new antenna.[89]

On 16 November 1999, Raytheon was selected by Boeing to provide the AESA radar. Under the agreement, Raytheon will develop an integrated AESA radar prototype. If this is successful, the Navy will award a contract for engineering and manufacturing development sometime in early 2001, with delivery of the AESA radar beginning in 2004. It is expected that the AESA radar will increase the F/A-18's air-to-air target detection range and tracking range; add higher-resolution air-to-ground mapping modes at longer ranges, enabling the aircraft to take full advantage of current and planned weapons; improve situational awareness in the cockpit; and significantly lower operating and support costs.

An integration test for F/A-18 Y2K compliance was conducted on the USS *John F. Kennedy* (CV-67) battle group from 22 June through 12 July 1999. The Y2K operational validation effort had begun with an extensive fast cruise workup conducted pierside in May through June 1999. During fast cruise, numerous systems were tested and validated, and crews were trained in preparation for the follow-on operational validation. Three squadrons of Lot X, XVI, XVII, and XVIII aircraft had their clocks set forward during the test. Mission-critical systems such as the mission computer, flight incident recorder and monitoring set, and armament control processor were tested for Y2K vulnerability and all passed.[90]

F/A-18D Two-Seater

Virtually all of the 40 F/A-18Bs were used exclusively for training, and the same is true of the first 31 F/A-18Ds, which are primarily assigned to Navy squadrons. The next 70 F/A-18Ds were configured as Night Attack Hornets, with the same changes as their single-seat counterparts. The prototype for the fully capable night attack F/A-18D was created by modifying the first F/A-18D (BuNo 163434), flying again at St. Louis on 6 May 1988. The first production Night Attack F/A-18D (BuNo 163986) was delivered to Pax River on 14 November 1989 and the first operational aircraft was delivered to VFMA(AW)-121 "Green Knights" on 11 May 1990. The last 60 aircraft (beginning with Lot XV) were completed as Night

The Marines have embraced the F/A-18D (BuNo 164702) as their primary all-weather attack air-craft now that the A-6E has been retired. This VMFA(AW)-533 example is typical. Note the AAS-38 pod on the air intake station. The MER on the outer wing pylon can carry only small practice bombs and is not cleared for operational weapons. This aircraft was photographed on 14 April 1993 at MCAS Beaufort. *(David F. Brown via the Mick Roth Collection)*

Attack F/A-18D(RC)s with a reconnaissance capability, making a total of 122 Night Attack two-seaters.[91]

The Night Attack F/A-18D is usually equipped with flight controls only in the front cock-pit. The flight officer seated in the aft cockpit is provided with two stationary hand controls, one on each side of the seat, to operate the weapons systems. However, in theory, the

A dark sky over NAS Lemoore provides a dramatic setting for this F/A-18D from VMFA(AW)-121 on 18 August 1996. *(Tom Chee via the Mick Roth Collection)*

Kuwait also ordered a version of the Night Attack F/A-18D, shown here at China Lake on 10 February 1993. These aircraft were marginally more capable than equivalent U.S. aircraft at the time since they were equipped with -402 enhanced performance engines (EPEs) and ALQ-165 advanced self-protection jammers (ASPJs). Later U.S. Hornets received similar equipment. *(Vance Vasquez via the Mick Roth Collection)*

F/A-18D may be reconfigured as a dual-control trainer. Originally this was seen as an 8 hour job involving removing the two side-console hand controllers and reinstalling the throttles and center control stick, as well as connecting the rudder pedals. In practice, the aft cockpit reconfiguration is labor-intensive and is seldom done. Fortunately, there is little operational requirement to perform the conversion on a routine basis.

Reconnaissance Hornets

In order to test the feasibility of a reconnaissance F/A-18 concept, McDonnell Douglas modified a single F/A-18A (BuNo 160775) with an internal sensor package in place of the 20-mm cannon. Sensors included a Fairchild-Weston KA-99 low/medium-altitude panoramic camera and a Honeywell AN/AAD-5 infrared linescanner. The aircraft could be converted back to the normal fighter configuration in only a few hours. The modified F-18(R)[92] flew for the first time on 15 August 1984. When this aircraft was removed from flight status and turned over to NASA as a source of spare parts, another F/A-18A (BuNo 161214) was converted to a similar configuration. Although not approved for production, the concept would find use by Australia on some of its Hornets, and also on the F/A-18D(RC) developed later.[93]

Next up was the "RF-18D" for the Marine Corps, although the intent was to be able to use the system on any F/A-18D wired to accept it, much like the TARPS on the F-14. The aircraft would carry an all-weather Loral UPD-4 side-looking high-resolution synthetic aperture radar reconnaissance pod on the centerline station, although there were also plans to use the improved Loral UPD-9 radar in the pod. Imagery could be display in the rear cockpit or transmitted in real-time via a self-contained data link. The pod, data link, and digital processing equipment was successfully demonstrated on an RF-4B during 1986. The final configuration was to be established in late 1987, with full deliveries scheduled to begin in 1990.

Even loaded with bombs, the F/A-18D is capable of climbing vertically. Notice that the canopy on the two-seater extends all the way back to the ALQ-165 blisters on top of the fuselage. Comparing this with a single-seater gives a good idea how much larger the canopy is. *(Boeing)*

Unfortunately, the system fell victim of defense budget constraints in 1988 and was canceled. The Marines were ordered to use a version of the Air Force advanced tactical air reconnaissance system (ATARS) pod instead.[94]

Development of ATARS was originally begun under an Air Force contract in May 1988 for use on a Teledyne Ryan drone, the F-16C, and later, Marine F/A-18s. The system was to be entirely pod-mounted except for a small control panel in the cockpit. Almost as soon as the requirements were defined, the Marines began incorporating provisions to carry the

The ATARS sensor pallet can be installed in place of the 20-mm cannon on most later F/A-18Ds. Although straightforward, the modification leaves some permanent indicators, even when the sensor pallet is removed. For instance, the ALQ-165 antenna blister normally mounted on the forward gear door is moved to the nose-wheel strut cover, and the ALR-65 low-band antennas are located on the forward gear door. *(U.S. Navy)*

ATARS pod into the Block 36 F/A-18D. The first F/A-18D (BuNo 164649) capable of carrying the ATARS was delivered on 14 February 1992, but unfortunately, no ATARS pods were delivered with it.

The program quickly ran into cost overruns and technical problems, forcing the Air Force to scale back its requirements, both in terms of imagery from the system and the total number of systems to be procured. The Air Force ultimately terminated the program in October 1993. The next month, however, the Navy received Congressional direction to resume development of ATARS as its tactical aerial reconnaissance system, and in January 1994 the Marine Corps assumed management of the ATARS program as lead service.[95]

The resulting AN/ZSD-1 advanced tactical air reconnaissance system—follow-on (ATARS-FO, but still normally called ATARS) is composed of an F/A-18D aircraft with an internally mounted nose sensor suite, a digital tape recorder, a digital data link pod, and a ground component called the joint service imagery processing station (JSIPS). The system provides Marine tactical commanders the organic capability to obtain imagery intelligence.

The F/A-18D(RC)[96] is an ATARS-equipped reconnaissance version of the two-seat F/A-18D that replaced the RF-4B Phantoms that served with VMFP-3 at MCAS El Toro. The squadron was redesignated VMFA(AW)-225 "Vagabonds" on 1 July 1991. The last 52 Night Attack F/A-18Ds (Lot XV and later) were completed to the (RC) configuration.

The normal M61A1 20-mm cannon is replaced by a pallet containing three primary sensors. The low-altitude electro-optical (LAEO) sensor operates between 200 and 3,000 feet with a 140° field of view and provides imagery from direct overflight of targets. The medium-altitude electro-optical (MAEO) sensor operates between 3,000 and 25,000 feet

The ATARS data link pod is carried on the fuselage centerline station. Other than eliminating the ability to use the 20-mm cannon, the ATARS modification has no adverse effect on the weapons capability of the F/A-18D. The use of this system gives the Marines an organic reconnaissance capability for the first time since the RF-4B Phantoms were retired in 1992. *(U.S. Navy)*

and has a 22° field of view. The MAEO is steerable, and can cover areas in a 220° swath at ranges of up to 5 nautical miles. These two sensors provide poststrike bomb damage assessments (BDA) and general imagery intelligence. The third sensor is the infrared line scanner (IRLS), which operates in either wide or narrow mode. Both modes operate between 200 and 25,000 feet and require overflight of the imaged area. The IRLS detects tactical targets such as recently operated vehicles or generators through their heat signature. Imagery obtained through the LAEO, MAEO, and IRLS sensors can be data-linked to a ground station if the aircraft is carrying an external ATARS data link pod (ADLP) on the centerline station.

The APG-73 RUG Phase II will provide the ability to create both reconnaissance strip maps and more detailed spot maps with very high resolutions. This system basically creates a radar-generated picture of any ground area of interest. The RUG II digital radar data can be recorded on the ATARS digital tape recorders, and may also be data-linked in near real time to the ground.

On 9 December 1998 a contract was awarded to produce six ATARS pallets and four data link pods. The contract also included aircraft modification kits, ancillary equipment, logistics support, contractor support, and training. It was the second low-rate initial production (LRIP) contract and was the first to procure the data link pods. The previous $50 million LRIP 1 contract procured four ATARS pallets. Total planned procurement is 31 ATARS sensor pallets, 31 aircraft conversion kits, 24 data link pods, and 58 RUG-II radar kits, although these requirements are not yet fully funded.[97]

Even before the ATARS had been subjected to formal testing, the Marines decided to conduct an operational evaluation by deploying it to Tazar, Hungary, as part the air campaign against Yugoslavia in May 1999. The system was deployed along with 24 F/A-18Ds belonging to VMFA(AW)-533 and VMFA(AW)-332 from MCAS Beaufort. Two aircraft were capable of using ATARS—both assigned to VMFA(AW)-332. The Marines deployed all three of their ATARS systems, but no data link pods were available until June 1999. Two ground stations where images can be processed were also deployed, and four imagery analysts were assigned to each squadron.[98]

One of the ATARS-capable F/A-18Ds had been used in development testing, while the other was hurriedly upgraded during early May 1999. During the 3-day-long upgrade, the cannon was removed, solid gun-door panels were replaced with panels equipped with sensor windows, various ECM antennas and the pitot tubes were relocated, and minor changes were made to the cooling system and wiring. Once an aircraft is modified, it takes about 1 hour to change between the cannon and the ATARS pallet. Even with the sensor suite installed, the aircraft retain all combat capabilities, except for the use of the cannon.[99]

"One potential use of the aircraft in Kosovo is strike coordination and reconnaissance," a Marine Corps official said before the aircraft deployed. In that scenario F/A-18Ds would spot a target, attack it, and later do their own bomb damage assessment using the reconnaissance system. The ATARS-equipped aircraft would carry a full load-out of air-to-ground and air-to-air weapons.

American Hornets in Desert Storm

In August 1990 Iraqi dictator Saddam Hussein invaded neighboring Kuwait. The United States responded by assembling a multinational coalition that amassed the largest armed force seen since World War II. Initially, troops and aircraft deployed to Saudi Arabia as part of Operation Desert Shield—intended to ensure Hussein did not try to overrun other Middle East countries. At the same time, diplomatic efforts resulted in an ultimatum from the United Nations demanding his unconditional withdrawal from occupied territories by 15 January 1991. Hussein defied the United Nations, and the next day the coalition responded with air attacks against Iraqi military positions in Kuwait and Iraq.

Operation Desert Storm was the first large-scale use of an entire new generation of American weapons, including the F/A-18. Department of Defense statistics show 88,500 tons of bombs were delivered in approximately 109,000 sorties flown by 2,800 fixed-wing aircraft. Of these sorties, the U.S. Air Force accounted for over 50 percent (using 1,300 aircraft), the U.S. Navy flew 16 percent (with 400 aircraft from six aircraft carriers), and the U.S. Marine Corps flew 9 percent (240 aircraft). The remaining 25 percent were flown by

The VFA-81 CAG was relatively understated, although the outside tail markings were "high-vis" instead of the normal subdued grays. The United States uses only the 330-gallon drop tanks because of deck-clearance constraints aboard aircraft carriers. Other nations use larger 480-gallon tanks, giving substantially more ferry range. *(Norris Graser via the Mick Roth Collection)*

All F/A-18C/Ds in the Fleet now carry the ALQ-165 blisters, although less than 100 systems were originally available from storage. Beginning in FY97, the Navy began purchasing new systems at a rate of about 30 per year, but it is unclear how many will ultimately be purchased. *(Mick Roth Collection)*

aircraft from other coalition forces. Just over half of the sorties were actual attack missions, while the remainder involved refueling, bomber escort, reconnaissance, or surveillance.

Nine Navy and seven Marine Corps Hornet squadrons participated in Operation Desert Storm. The Navy brought 106 F/A-18A/Cs on six aircraft carriers, while the Marines operated 72 F/A-18A/Cs and 12 F/A-18Ds from land bases.

The first four Navy F/A-18A squadrons in the area were VFA-25 "Fist of the Fleet" (NK) and VFA-113 "Stingers" (NK) aboard the USS *Independence* (CV-62), and VFA-131 "Wildcats" (AG) and VFA-136 "Knighthawks" (AG) aboard the USS *Eisenhower* (CVN-69).

When these squadrons rotated home, they were replaced by three F/A-18A squadrons—

When the aircraft is parked, the flaps and ailerons are normally full down, since the hydraulic system is unpressurized. It is unusual, however, for the inflight refueling probe to be extended. This F/A-18C (BuNo 164668) was photographed in November 1994. *(Mick Roth Collection)*

VFA-151 "Vigilantes" (tailcode NM), VFA-192 "Golden Dragons" (NF), and VFA-195 "Dam Busters" (NF) aboard USS *Midway* (CV-41); two F/A-18A squadrons—VFA-15 "Valions" (AJ) and VFA-87 "Golden Warriors" (AJ) aboard USS *Theodore Roosevelt* (CVN-71); and two F/A-18C squadrons—VFA-81 "Sunliners" (AA) and VFA-83 "Rampagers" (AA) aboard USS *Saratoga* (CV-60). USS *America* (CV-66), with her two F/A-18C squadrons—VFA-82 "Marauders" (AB) and VFA-86 "Sidewinders" (AB)—subsequently deployed to the region. Other non-Hornet air assets on board USS *John F. Kennedy* (CV-67) and USS *Ranger* (CV-61) were also on station.

The Marine Corps deployed F/A-18As with VMFA-314 "Black Knights" (tailcode VW), VMFA-333 "Shamrocks" (DN), and VMFA-451 "Warlords" (VM). Marine F/A-18Cs, all from MAG-24, flew with VMFA-212 "Lancers" (WD), VMFA-232 "Red Devils" (WT), and VMFA-235 "Death Angels" (DB). Marine F/A-18Ds from MAG-11 operated with VMFA(AW)-121 "Green Knights" (VK).

The F/A-18 proved its versatility by shooting down enemy fighters and subsequently bombing enemy targets using the same aircraft on the same mission. According to published figures, the F/A-18 availability rate during Operation Desert Storm averaged 90 percent and peaked at 95 percent. Further analysis reveals, however, that this reflected devoting most of the Navy's resources (spare parts, etc.) to sustaining only half the Hornet force.

There was a clear difference between Navy and Marine F/A-18 tasking during Operation Desert Storm. Navy missions fell equally into three categories—strike missions (36 percent), fleet defense (30 percent), and support (34 percent). Marine missions were predominately direct combat sorties (84 percent), augmented by a small number of support sorties (16 percent). F/A-18s flew 4,551 combat sorties with 10 aircraft sustaining combat damage, equating to a casualty rate per strike of 0.22 percent. Two of these 10 aircraft were lost in combat, while the other eight managed to return to a friendly base:[100]

Date	Result	Unit	Cause
17 January 1991	Loss	VFA-81	MiG-25PD
5 February 1991	Loss	USN	Unknown
9 February 1991	Damage	USMC	IR-SAM
21 February 1991	Damage	VMFA-314	IR-SAM
21 February 1991	Damage	VMFA-121	IR-SAM
21 February 1991	Damage	VMFA-333	IR-SAM
22 February 1991	Damage	VMFA-451	IR-SAM
24 February 1991	Damage	VMFA-314	IR-SAM
24 February 1991	Damage	VMFA-314	IR-SAM
24 February 1991	Damage	USMC	Small arms

One of the F/A-18s was hit in both engines, yet flew 125 miles to recover at its home base. The aircraft was repaired and returned to service within a few days.[101] In addition, two Marine F/A-18Cs were involved in a midair collision over Saudi Arabia on 9 March, and fortunately both pilots ejected safely.

A large number of early F/A-18 missions involved escorting large strike packages of A-6Es supported by EA-6Bs into southern Iraq. Along with Air Force F-15s and Navy F-14s, the F/A-18s provided combat air patrols (CAP), and quickly established air superiority over

An F/A-18C from VFA-94 snags the #1 wire aboard USS *Carl Vinson* (CVN-70) during operations off Puget Sound in October 1997. The aircraft is too low and the right wing is down—not a pretty landing. Note the A-6E parked to the side. *(U.S. Navy/PMA Wesley Barnard)*

A catapult safety officer monitors the flight line as an F/A-18C launches from one of the bow catapults aboard USS *Nimitz* (CVN-68) during a 1997 diversion to the Persian Gulf. The Hornet is carrying an AGM-88 HARM missile. *(U.S. Navy)*

The VFA-86 F/A-18C (BuNo 163439) of Commander Zack "Sheets" May displays missions marks from the Gulf war. Four Shrike symbols and 15 bomb symbols indicate the number of missions flown. The photo was taken aboard USS *America* in Naples during January 1992. *(Alfredo Maglione via the Terry Panopalis Collection)*

the aircraft carriers. During attacks on Silkworm antiship missile sites, F/A-18s used AGM-142 Walleyes, AGM-84E SLAMs, and Mk 80-series iron bombs. Marine F/A-18s also dropped snake-retarded versions of the Mk 80s on Iraqi troop positions in Kuwait.

F/A-18s delivered 5,513 tons of ordnance and averaged 1.2 daily sorties (about the same as the A-6E). However, the Hornet delivered an average of only 0.74 ton of munitions per day, the lowest of any aircraft. By comparison, the A-6E averaged 1.16 tons of munitions per day. The Hornet's lack of an autonomous laser designator was considered a serious short-coming, one which contributed to the unflattering tonnage figure. The addition of a laser designator to the AAS-38A/B pod has largely cured this problem.

The F/A-18 delivered one guided weapon for every 30 unguided weapons delivered—368 guided versus 11,179 unguided. By comparison, the F-15E had a ratio of 1:8 (1,669 guided; 14,089 unguided). Although the F/A-18 did deliver cluster bombs and precision-guided munitions, the Navy relied predominantly on the night-capable A-6E for laser-guided bomb delivery and used the F/A-18 primarily to deliver iron bombs, usually the 1,000-pound Mk 83. In fact, the largest amount of bombs delivered on a single tactical mission—90,000 pounds of Mk83s—was dropped when all 18 F/A-18s aboard USS *Saratoga* attacked Iraqi positions in Kuwait.[102]

Although nearly 85 percent of downed Iraqi aircraft during Operation Desert Storm were accounted for by Air Force F-15Cs, the F/A-18 is credited with two air-to-air kills. Four Hornets from VFA-81 "Sunliners" off the USS *Saratoga* were on their way to a target on 17 January when two of them were engaged by Iraqi MiG-21 Fishbeds. The Hornets were in contact with an E-2C Hawkeye (159107/AA-600) at the time. Lieutenant Commander Mark Fox was flying an F/A-18C (BuNo 163508/AA-401) and fired an AIM-9 Sidewinder at a second MiG. The Sidewinder was using a smokeless motor, and Fox lost track of the missile and assumed it had malfunctioned. Fox then fired an AIM-7 Sparrow. Both missiles hit the MiG. Lieutenant Nick Mongillo was flying another F/A-18C (BuNo

A careful examination will show a small MiG-21 silhouette on the nose of this F/A-18C (BuNo 163502) flown by Lieutenant Nick Mongillo from VFA-81. He and Lieutenant Commander Mark Fox each scored an aerial victory on 17 January 1991. The photo was taken aboard USS *America* (CV-66) in Naples during January 1992. *(Alfredo Maglione via the Terry Panopalis Collection)*

163502/AA-410) and fired a single AIM-9, destroying the second MiG. Neither pilot jettisoned his bombs, and both continued their mission through SAM and AAA fire. The strike force dropped their bombs on their original target, an airfield in western Iraq, and returned to their base.

From Lieutenant Commander Fox's Unit Mission Report:

> We crossed the Iraqi border in an offset battle box formation to maintain the best lookout possible. As the strike developed, the volume and intensity of communications over the strike frequency increased. Bandit calls from the E-2 to our other strike group crowded in to my mind as I plotted where those bandits should be relative to our position. A call from the E-2 clearly intended for the Hornet strikers finally registered: "Bandits on your nose, 15 miles!" I immediately selected Sidewinder and obtained a radar lock on a head-on, supersonic Iraqi MiG-21. I fired a Sidewinder and lost sight of it while concentrating on watching the MiG. Thinking the Sidewinder wasn't tracking, I selected Sparrow and fired. A few seconds after the Sparrow left the rail, the Sidewinder impacted the MiG-21 with a bright flash and puff of black smoke. Trailing flame, the MiG was hit seconds later by the Sparrow and began a pronounced deceleration and descent. As the flaming MiG passed below me, I rocked up on my left wing to watch him go by. Another F/A-18 pilot killed the MiG's wingman with a Sparrow shot [a Sidewinder, as later reports revealed] only seconds after my missiles impacted the lead MiG. . . . After the tactical activity associated with bagging a MiG while entering a high threat target area, the dive bombing run on our primary target was effortless. Visible below me were numerous muzzle flashes, dust and smoke from gun emplacements, a light carpet of AAA bursts and several corkscrew streaks of handheld SAMs being fired. I glanced back at the target just in time to see my four 2,000 pound bombs explode on the hangar. Our division quickly reformed off target without incident and beat a hasty retreat south of the border. Our relief in having successfully completed the strike without loss to ourselves was overwhelming.

On the first night of Operation Desert Storm, Lieutenant Commander Michael "Spike" Speicher from VFA-81 was flying an F/A-18C (BuNo 163484/AA-403) when he was shot down 33 miles southeast of Baghdad. Although some initial sources attributed his loss to ground fire (specifically a Soviet-made SA-6 SAM), most accounts attribute the downing to an Iraqi MiG-25PD Foxbat. Speicher was the only U.S. casualty unaccounted for in the Gulf war, and was listed as missing until 1996 when he was declared dead.

Operation Deliberate Force

Marine F/A-18Ds flew more than 100 strike sorties against Serbian military targets beginning in August 1995. Many of these operations were SEAD missions using GBU-16 1,000-pound laser-guided bombs and AGM-88 HARM antiradiation missiles. Like many modern military operations, the use of laser-guided bombs was absolutely essential since the conflict was being waged against the government, not the people, of Serbia. Collateral damage was to be minimized at all costs, ruling out the possibility of using "dumb" bombs from strategic bombers.

Operation Allied Force

In late 1998 NATO became increasingly worried about the actions of Slobodan Milosevic in Kosovo. The U.S. Air Force formed four air expeditionary wings on 11 October 1998 to help simplify lines of command and control should a NATO-led force be directed on Kosovo. These included the 16th Air Expeditionary Wing and 31st Air Expeditionary Fighter Wing at Aviano Air Base (AB), Italy, the 86th Air Expeditionary Airlift Wing at Ramstein AB, Germany, and the 100th Expeditionary Air Refueling Wing at RAF Mildenhall, United Kingdom.

Hornets under the command of these air wings included EF-18s from Spain, CF-18s from Canada, and U.S. Marine Corps F/A-18C/Ds. U.S. Navy F/A-18s were stationed aboard the USS *Theodore Roosevelt* (CVN-71), which deployed from the United States on 26 March 1999 for a 6-month cruise in the Mediterranean. As of 6 May 1999, the Department of Defense announced that the total number of U.S. aircraft committed to Operation Allied Force was 639, along with 277 from 13 other North Atlantic Treaty Organization (NATO) nations. Although the United States had deployed a large percentage of aircraft to the region, actual sorties averaged 58 percent American, and 42 percent other NATO countries.

In support of NATO Operation Allied Force, the Canadian Operation Echo began in June 1998 with the movement of six CF-18s and approximately 130 Canadian Forces members to Aviano AB in readiness for possible NATO action in the region, and to help enforce the no-fly zone over Bosnia. All of the CF-18s were fitted with air-to-air missiles and were capable of dropping nonprecision and precision-guided weapons for ground attack. On 23 March 1999, NATO Secretary General Javier Solana ordered the Supreme Allied Commander in Europe (SACEUR), General Wesley Clark, to begin air operations in Yugoslavia. This decision was taken after the failure of diplomatic negotiations. As the air campaign began, the Canadian contingent in Aviano was made up mostly of personnel and aircraft from 3 Wing at Bagotville under the command of Colonel Dwight Davies. On 24 March the first wave of attacks was carried out by NATO aircraft, including CF-18s. On 30 March Canadian Defence Minister Art Eggleton announced that six more CF-18s were being sent to Aviano in response to a request from NATO's military command, thus bringing the total number of aircraft to 12. These six aircraft arrived at Aviano on 2 April, and six additional aircraft follow on 27 April. The Canadian Forces aircraft flew an average of 16 sorties per day, with a maximum surge of 20 sorties on 11 May.[103]

At the beginning of May, the NATO nations were conducting around 270 combat sorties per day over Kosovo and Serbia in an attempt to defeat the ethnic cleansing reportedly underway by Milosevic's troops. By the end of May this was up to 1,000 sorties per day. Part of the increase was possible because on 20 May the Marines deployed 24 Night Attack F/A-18Ds from VMFA(AW)-332 and VMFA(AW)-533 to the former Warsaw-pact air base at Taszar, Hungary, becoming operational on 28 May. The Marine F/A-18Ds performed a wide

Lieutenant Ron Candiloro's Hornet creates a shock wave as he breaks the sound barrier on 7 July 1999. The shock wave is visible as a large cloud of condensation formed by the cooling of the air. A smaller shock wave can be seen forming on top of the canopy. It is possible for a skilled pilot to work the plane's throttle to move the shock wave forward or aft. Candiloro was assigned to VFA-151 deployed aboard the USS *Constellation*. *(U.S. Navy photo by Ensign John Gay)*

variety of missions during Allied Force, including deep air strikes, forward air control (FAC), close air support, aerial reconnaissance, strike coordination, and antiair warfare. Typical FAC missions used night vision goggles and lasted for over 6 hours, including four aerial refuelings. Deep air strikes generally used 1,000-pound laser-guided bombs, and were frequently challenged by an unexpectedly robust SA-3 SAM capability. Conversely, during the time the two Marine squadrons were on station, not a single Serbian MiG challenged them. Nevertheless, the F/A-18Ds always carried two AIM-9s and at least one AIM-120 just in case. During the campaign the two Marine squadrons expended 304,000 pounds of ordnance.[104]

However, it is the Spanish Air Force that holds the record for longevity at Aviano. EF-18s of the Icarus Detachment arrived at Aviano in December 1994 in support of Operation Deny Flight. They never left, accumulating more than 14,000 flight hours during three NATO air operations based from Aviano. During Operation Allied Force, the detachment included six EF-18s from *Escuadron* 122 plus a KC-130 tanker from *Escuadron* 312. The normal sortie rate for the detachment was two EF-18 sorties in the morning, two in the afternoon, and two at night—a testimony to the mission-capable rate of the Hornet. Ironically, the KC-130 tanker refueled non-Spanish aircraft much more frequently than it did the EF-18s, which usually tanked from NATO tankers. It all depended on what aircraft was where, not what nationality it belonged to.[105]

Foreign Hornets: Widely Exported

Historically, Navy aircraft have not done well in the export market. In recent times, France and the Philippines purchased a small number of Vought F-8 Crusaders, Iran purchased 80 Grumman F-14A Tomcats, and the United Kingdom purchased modified versions of the F-4. But the vast majority of foreign F-4 and A-7 sales were of the Air Force variants. In recent memory, only the McDonnell Douglas A-4 Skyhawk has provided any significant sales opportunities for a Navy aircraft.[106]

The export Hornets are generally similar to the U.S. models being produced at the time. Since none of the countries currently operate aircraft carriers, all export customers have elected to delete the automatic carrier landing system (ACLS), and Australia deleted the catapult attachments on the nose landing gear. Otherwise, except for variations in electronics and software, all Hornets are generally alike. With the exception of Canada, all export customers have ordered their aircraft through the U.S. Navy as part of the U.S. Foreign Military Sales (FMS) program. Under this program the Navy acts as a purchasing manager for the foreign customer, but is prohibited by law from making a profit (or incurring a loss) on the sale. Unlike the Air Force, which generally assigns a USAF serial number to all FMS aircraft, the Navy tracks the export aircraft via the customer's serial number, not a BuNo. Enjoying a unique relationship with the United States, Canada placed its order directly with McDonnell Douglas.

Canada

In March 1977 the Canadian government authorized the Department of National Defence to acquire a New Fighter Aircraft (NFA) to replace its CF-101 Voodoo, CF-5 Freedom Fighter, and the CF-104 Starfighter types. Very rapidly the primary competitors became the F-16 and the F/A-18, although Northrop attempted to sell the F-18L. In fact, Canadian pilots were very enthusiastic when they evaluated the YF-17 while it was serving as the F-18L demonstrator.

On 10 April 1980 Canada announced that the F/A-18 had been selected largely because of its two engines, the Canadians believing this was more suitable to Canada's environment,

The No. 410 OCU Squadron painted this CF-18A (188746) in celebration of the fiftieth anniversary of the Royal Canadian Air Force (RCAF) roundel. The large maple leaf on the tail is particularly striking. The CF-18s are essentially identical to the U.S. F/A-18A/B, although they are slowly being brought up to C/D standard. Note the large searchlight on the forward fuselage under the tip of the LEX. *(Mark Munzel)*

which includes operations over large isolated regions including the Arctic. The order was for 113 single-seaters and 24 two-seaters, with options being taken for 20 additional aircraft. The initial plan was to order 11 more aircraft, but the option expired unexercised on 1 April 1985. The original contract was subsequently modified to 98 single-seaters and 40 two-seaters for C$2,340 million in 1977 dollars. This represented the largest single defense contract in Canadian history.[107]

The major external modification is the addition of a 600,000-candlepower spotlight fitted on the left side of the forward fuselage (under the tip of the LEX) to enable night identification of other aircraft. The maritime rescue package on the NACES ejection seat was exchanged for cold-weather survival equipment, and the ACLS was replaced by a civilian instrument landing system (ILS). In addition, the Canadian aircraft have provisions to carry LAU-5003 rocket pods (containing 19 Bristol Aerospace CRV-7 2.75-inch unguided rockets) and BL-755 cluster bombs. The aircraft also carry three 480-gallon drop tanks instead of the normal 330-gallon units, and were fitted with ALQ-162 jammers to complement the normal ALQ-126B jammers and ALR-67 RWR.

The aircraft is designated CF-18 (single seat) and CF-18B (two seats) in Canadian Forces service.[108] The two-seater was initially designated CF-18D (for *dual*) following previous Canadian practice, but this was subsequently changed to CF-18B to avoid confusion with the F/A-18D.[109]

The first production CF-18 made its maiden flight at St. Louis on 29 July 1982, and the first two aircraft (188901 and 188902) were delivered on 25 October. The CF-18As were assigned Canadian military serials 188701 through 188798, and the 40 two-seat CF-18Bs were assigned 188901 through 188940. All the Canadian Hornets were initially delivered to No. 410 "Cougars" Operational Training Unit for acceptance testing prior to allocation to operational squadrons, with the last being delivered to Cold Lake AFB on 28 September 1988.[110]

Much more so than the Americans, Canadian Forces allow the squadrons to paint aircraft in special markings for a variety of occasions. This CF-18A (188730) from No. 433 Squadron is celebrating the 130th Anniversary of Canada (1867–1997). Believe it or not, the snarling animal on the top of the vertical is a porcupine, the squadron's namesake. *(David F. Brown via the Mick Roth Collection)*

The CF-18 has served with No. 416 "Lynxes" and No. 441 "Silver Foxes" Squadrons at Cold Lake, Alberta; No. 425 "Alouettes" and No. 433 "Porcupines" Squadrons at Bagotville, Quebec; and No. 409 "Nighthawks," No. 421 "Red Indians," and No. 439 "Tigers" Squadrons at Baden-Soellingen in Germany.

Canada sent 18 aircraft from No. 409 Squadron to Doha, Qatar, on 7 October 1990 to participate in Operation Desert Shield. From this base, dubbed "Canada Dry," CF-18s flew over 1,110 CAP and training sorties in the months leading up to the Gulf War. No. 409 Squadron was supplemented in mid-December 1990 by 26 CF-18s from Nos. 439 and 416 Squadrons. The resulting Canadian Air Task Group Middle East (CATGME) also included a Boeing 707 tanker. The CF-18s conducted CAP missions over Saudi Arabia and the Gulf, and also escorted strike packages consisting of Royal Air Force (RAF) Buccaneers and Tornados and U.S. F-16s.

This CF-18A from No. 425 Squadron is in fairly standard markings for the type. Since Canada has two official languages (English and French), the writing on either side of the roundel says "Armed Forces" and "Forces Armée." All markings are usually in a subdued gray paint, although it is not uncommon to see a full color roundel or fin flash. *(Mick Roth Collection)*

Another special was this No. 410 OCU Squadron CF-18A (188710) with a fighter pilot on the vertical. The crests of all five CF-18 squadrons are shown, along with the caption "For the Defence of Canada/Pour la Defense du Canada." Although quite striking, the markings were basically gray with a yellow and red leading edge. *(Mike Reyno/Skytech Images)*

The CF-18s dropped 500-pound Mk 82 iron bombs and also used laser-guided bombs in limited numbers. Since the CF-18s lacked laser designators, other aircraft—usually U.S. Navy A-6Es—illuminated the targets. For example, on 27 January 1991, an A-6E that had already successfully expended its two 500-pound laser-guided bombs on two Iraqi patrol boats illuminated a third patrol boat for a CF-18. Two other CF-18s—piloted by Major David "DW" Kendall and Captain Steve "Hillbilly" Hill—were called in to finish the job. Although pulled off CAP duty and not configured with air-to-ground ordnance, they attacked and severely damaged the last of the four patrol boats (an Exocet-equipped TCN-45) with 20-mm cannon fire. Kendall also fired an AIM-7 Sparrow, which impacted the water near the boat. This was the first time Canadian forces had fired weapons in anger since the Korean War. In total, the CF-18s logged 5,730 flight hours with no combat losses or damage, and the 707 tanker flew an additional 306 hours.

Following its participation in Operation Desert Storm, No. 409 Squadron was disbanded in 1991, turning over some of its aircraft to Nos. 421 and 439 Squadrons. As part of a general stand-down in Europe, No. 421 Squadron disbanded in June 1992, and No. 439 stood down in December 1992. Baden-Soellingen was closed in 1994 after the end of the Cold War. This left only four active duty squadrons—Nos. 416, 425, 433, and 441 Squadrons—plus No. 410 TF(OT)S training squadron at Cold Lake. Two of these squadrons are on notice for a quick return to Europe if needed, while the other two are assigned to support maritime operations. Following the disestablishment of the European-based CF-18 squadrons, some of their aircraft were redistributed to the four Canadian-based squadrons, and the remainder were placed in flyable storage. At the end of 1994, only 72 CF-18s remained in operational squadrons, with the remainder serving with the training unit or being held in storage.

In 1995 the Canadian Forces Air Command announced that a further 12 CF-18s would be withdrawn from active service and placed into ready reserve storage. This left only 60 CF-18s in four operational squadrons, each with 15 rather than 18 aircraft, and 27 CF-18s with No. 410 TF(OT)S.[111] Even given their reduced numbers, CF-18s continue to provide "flexible alert" for the North American Air Defense Command (NORAD). The four operational

Two CF-18As and a CF-18B of No. 441 Squadron. The Canadians are the only Hornet operator to universally use the "false canopy" on the bottom of the forward fuselage in an attempt to disguise the attitude of the aircraft. *(Mike Reyno/Skytech Images)*

squadrons rotate the duty to ensure CF-18s are available 24 hours per day, 7 days per week, at the deployed operating bases at Goose Bay and Comox, as well as periodically at the forward operating locations at Inuvik, Yellowknife, Rankin Inlet, and Iqaluit.[112]

Canada initially envisioned flying the CF-18 until 2003, but because of fiscal realities it has been realized that the aircraft will need to be maintained in service much longer. To ensure the aircraft's future viability, Canada has initiated the CF-18 Incremental Modernization Project (IMP) to extend their operational life until 2017 to 2020. Lieutenant Colonel Dave Burt, project director for the IMP, said the upgrade will provide increased capability for the CF-18, interoperability to conduct operations with NATO and NORAD nations, and supportability. "Supportability is our biggest concern right now. The CF-18 fleet is encountering parts obsolescence and supportability issues, and its operational capability could markedly degrade in the near future," according to Burt. Some suppliers have said they will cease supporting systems used on the CF-18 as early as the year 2000, making it increasingly difficult to support some systems as the number of spare parts diminishes.[113]

Captain Steve Nierlich, a CF-18 instructor pilot[114] with No. 410 TF(OT)S said: "Most NATO air forces are using equipment such as HaveQuick radios permitting pilots to talk to ground control intercept (GCI) and AWACS [airborne warning and control system] controllers without fear of being jammed themselves by the opposition. The CF-18, however, does not have this capability—which means we have to work autonomously since we can't talk to anyone and no one can talk to us in a high-threat scenario without being jammed. The same holds true if I have a radar contact coming at me. Since the CF-18 doesn't have an identify friend or foe (IFF) interrogator, combined with the fact that I can't talk to GCI or AWACS to confirm what's coming at me, I have to wait until I have a visual identification because no one can tell us if he's a bad guy or not. The risk of being shot at, or shot down for that matter, has gone up exponentially."[115]

The CF-18 fleet presently has an average of 3,200 airframe hours with a life expectancy of 6,000 hours originally projected by McDonnell Douglas. The structural life will be extended to 2017 to 2020 by the International Follow-On Structural Program being planned with

This No. 441 Squadron CF-18A (188781) was painted with invasion stripes to commemorate the fiftieth anniversary of the D-Day invasion in June 1944. A flyover involving American, British, French, Canadian, Belgian, Greek, Dutch, Norwegian, Czech, and Slovak combat aircraft took place over Portsmouth in 5 June 1994, and over France the following day. Altogether, 11 CF-18s were involved. *(Mike Reyno/Skytech Images)*

Australia. The CF-18 is the only Canadian Forces aircraft that is managed by its fatigue life and not airframe hours. Each aircraft is equipped with load-monitoring sensors that measure the amount of structural fatigue encountered during each sortie. Reports are printed monthly on each CF-18 to determine the amount of fatigue accumulated, and an analysis of this data determines what mission types each individual aircraft will be used for in order to maximize its life expectancy.

The CF-18 IMP incorporates 13 separate projects that will be integrated and installed in three phases to minimize the cost of the overall program. The first phase includes new mission software, ARC-210 HaveQuick radios, GPS/INS (inertial navigation system), and XN-10 mission computers. In the second phase, the existing APG-65 will be replaced by the APG-73, which Canada helped develop. New digital displays and an IFF interrogator will also be installed. The final phase includes a modernized ECM suite, effectively bringing the aircraft up to C/D standard. The first CF-18 could enter IMP in 2001, with the first fully upgraded CF-18 that has gone through all three phases ready in 2007. The last aircraft will be completed in 2009.

Although the Canadian government has endorsed the C$1,200 million program, only C$60 million has been approved for the first phase, with the Air Force having to seek funding for each follow-on item separately. One option being explored to help offset the program's costs is the possible sale of some CF-18s now in storage. "As it stands right now we anticipate on modernizing up to 100 CF-18s, so we have examined selling or leasing those CF-18s considered 'surplus' to help fund the program," Burt said. "They could also be used for spare parts to help alleviate our supportability issues. But these are ideas being considered and it will be quite some time before any decisions will be made regarding those aircraft that will not be upgraded."

Separately, during 1998 Canada upgraded to the AAS-38A LDT/R pod, allowing the Hornets to designate their own targets for laser-guided bombs. Interestingly, the normal load for the CF-18s is an AAS-38A pod on one fuselage station, and an AIM-7 Sparrow on the other.[116]

One of the more spectacular, and more publicized, specials is this CF-18A (188718) from No. 410 TF(OT) Squadron. The aircraft is painted in light and dark blue trim with old-style RCAF roundels, the RCAF tartan, and the king and queen crowns on each vertical stabilizer. The aircraft appeared at air shows around North America during 1999 to celebrate the seventy-fifth anniversary of the Royal Canadian Air Force. *(Mike Reyno/Skytech Images)*

In otherwise standard subdued markings, this CF-18A (188786) from No. 433 Squadron wears a red and white fin flash. Like all A/B models, the CF-18s have been fitted with a variety of doublers on the vertical stabilizer attachment points to correct a structural fatigue problem caused by turbulence flowing over the LEX impacting the verticals. *(Mike Reyno/Skytech Images)*

Australia

On 20 October 1981 the Royal Australian Air Force (RAAF) announced that it had selected the F/A-18 as the replacement for their Dassault Mirage IIIOs. The RAAF order was for 57 single-seaters (A21-1 to A21-57) and 18 two-seaters (A21-101 to A-21-118).[117] The single-seater is usually listed as AF-18A and the two-seater as AF-18B, with the "A" standing for "Australia," although these are not official Department of Defense (DoD) designations.

Like many recent arms purchases, the Australian order included a complex financial offset arrangement, with 40 percent of the components being manufactured in Australia. McDonnell Douglas was responsible for the manufacture of the first two aircraft, with the Government Aircraft Factory (later renamed Aero-Space Technologies of Australia, or ASTA) at Avalon, Victoria, assembling the remainder from parts supplied by both U.S. and Australian suppliers. Australian responsibilities included the manufacture of forward fuselage, trailing-edge flaps, radome assemblies, transparencies, and final assembly. Dunlop Aviation Australia was to make the wheels and brakes as well as the airspeed indicator. Some software was provided by Computer Sciences Australia, and electronic components were provided by Morris Productions, Philips, Thorn EMI Electronics Australia, and Standard Telephones and Cables. The F404 turbofans were built under license by the Commonwealth Aircraft Corporation, with the APG-65 radar and other avionics being supplied by British Aerospace Australia, Ltd.

McDonnell Douglas shipped components for the first two Australian-assembled AF-18Bs (A21-103 and A21-104) to Avalon inside a C-5A, arriving on 6 June 1984. The two AF-18Bs manufactured by McDonnell Douglas in St. Louis were handed over on 29 October 1984, although both aircraft were retained at St. Louis for training and flight testing until 17 May 1985, when they were flown to RAAF Williamtown. The aircraft refueled numerous times from U.S. Air Force KC-10 tankers during the 15-hour direct flight, which was led by the No. 2 Operational Conversion Unit (OCU) Commanding Officer, Wing Commander Brian Robinson.[118]

The remaining aircraft on the order were all assembled in Australia, with the first (A21-103) making its maiden flight on 26 February 1985 with McDonnell Douglas test pilot Rudi Haug at the controls. The aircraft reached a speed of Mach 1.6 at 40,000 feet on this

This AF-18A (A21-40) from the Royal Australian Air Force No. 77 Squadron shows a small vapor trail coming off the LEX at a high angle of attack. The aircraft is still wearing high-visibility roundels and fin flash. The RAAF AF-18s are externally identical to the U.S. A/B models. *(RAAF/LAC Jason Freeman via Chris Hake)*

flight. This aircraft was accepted by the RAAF on 30 April 1985, formally handed over on 4 May, and then delivered to No. 2 OCU on 17 May. On 25 November the first single-seater (A21-1) was accepted.[119]

The first 14 aircraft (A21-1 to A21-7 and A21-101 to A21-107) were all delivered to No. 2 OCU for Hornet instructor training. The first RAAF Hornet conversion course began on 19 August 1985, and the AF-18 simulator was installed later that year. AF-18s currently serve with Nos. 3 and 77 Squadrons at RAAF Williamtown and No. 75 Squadron at RAAF Tindal in the Northern Territory. Each squadron is usually allocated 1 or 2 two-seaters, with the rest serving with No. 2 OCU.[120]

Four AF-18As from No. 3 Squadron with Mk 82 selectables. The RAAF AF-18As, much like the Canadian CF-18s, can carry the same weapons as their American counterparts, and also a small variety of British weapons [CVR-7 folding-fin aircraft rockets (FFARs), BL 755 cluster bombs, etc.]. *(RAAF/LAC Jason Freeman via Chris Hake)*

Two AF-18As (A21-12 and A21-18) from No. 3 Squadron maneuver off the coast of New South Wales. The RAAF does not carry national insignia on top of the wings, and the two roundels on the bottom of the wings are oriented toward the centerline of the aircraft, not the direction of flight. Without wing pylons and drop tanks, the Hornet is a very clean-looking aircraft. *(RAAF/LAC Jason Freeman via Chris Hake)*

The last AF-18B (A21-118) was delivered on 15 December 1988, and the final RAAF Hornet (A21-57) was handed over on 12 May 1990. The aircraft was delivered to Williamtown on 14 May, flown by Squadron Leader Ron Haack, and was the subject of a formal delivery ceremony at RAAF Fairburn on 16 May.

Two AF-18As and two AF-18Bs have been lost in accidents. A21-104 was lost in November 1987, and A21-41 was lost in a midair collision with A21-29 (which landed at Tindal) in August 1990. A21-41 was lost in June 1991, and A21-106 was lost in May 1992.

The AF-18 deletes the ACLS, and is equipped with conventional instrument landing system/very high frequency omnidirectional range (ILS/VOR) equipment, fatigue recorder, and an additional high-frequency radio for long-range communications. The catapult launch bar was not fitted to the AF-18 since shipboard operation was not required. However, the removal of the launch bar upset the nose-strut dynamics and the result was a nose-wheel shimmy ranging in intensity from annoying to frightening. A non-functioning launch bar subsequently was fitted in order to damp the nosewheel oscillations. Australian Hornets are capable of launching the AGM-65 Maverick and the AGM-84 Harpoon. The aircraft are equipped with an ALR-67(V) radar warning receiver (RWR), ALQ-126B jammer, and ALE-39 dispenser. The RAAF has also performed the software and wiring modifications necessary to carry the AIM-120 advanced medium-range air-to-air missile (AMRAAM).

Twenty-three of the AF-18s have provisions to replace the M61 20-mm cannon with a reconnaissance sensor pallet. Available sensors include the KA-56 3-inch panoramic camera, KS-87 6-inch side oblique camera, KA-93 24-inch sector panoramic camera, and the KS-87 12-inch split vertical camera. This is not the same pallet as the AN/ZSD-1 ATARS-FO pallet carried on Marine F/A-18D(RC)s.

The top hat markings of this No. 75 Squadron AF-18A (A21-29) adorn the commanding officer's aircraft. Australia's Hornets will soon sprout additional ECM antenna blisters as the electronic systems are upgraded. Also to be upgraded are the radar and computers, although for the time being, Australia has elected not to procure the -402 EPE engine. *(RAAF/LAC Peter Gammie via Chris Hake)*

An AF-18 (A21-5) from No. 77 Squadron flies over one of the most photographed locations in the world—Sydney Harbour and the Opera House. Australia elected not to equip its aircraft with the catapult launch bar on the nose landing gear strut, leading to a shimmy while taxing under certain conditions. A "dummy" launch bar is being installed that will correct this condition. *(RAAF/LAC Peter Gammie via Chris Hake)*

This AF-18A (A21-32) visited China Lake to perform weapons evaluations. The bombs in this photo were yellow, green, and red, providing a quick reference when videotaped. Note the cameras on the wingtips and under the rear fuselage. *(Tony Landis Collection)*

To allow the AF-18 to remain in service with the RAAF until 2015, a $710 million Hornet Upgrade (HUG) is under way, conducted by Boeing Australia. The HUG will incorporate some systems common to production F/A-18C/Ds, and several in development for future aircraft. The first phase of the upgrade, worth $145 million, includes adding ARC-210 Have-Quick radios, XN8+ mission computers, improved IFF equipment, additional combat maneuvering instrumentation wiring, ECM software upgrades, a Global Positioning System (GPS), and a sixth data bus to allow integration of future smart weapons. The $116 million second phase will add the APG-73 radar, Link 16, a color display tactical map system, and the joint helmet-mounted cueing system (JHMCS) for weapon sighting.[121] The JHMCS will be fully integrated with the AIM-132 ASRAAM. A fully funded third phase is expected to bring the AF-18 ECM suite up to the same standard as current C/Ds with the ALQ-165 ASPJ and ALE-47 dispensers. One exception may be the substitution of the Australian-developed ALR-2002 radar warning receiver instead of the ALR-67(V)3 RWR. RAAF hornets are also being modified to enable NVG (ANVIS-6) compatibility, complementing the current AAS-38 FLIR.[122]

Spain

In December 1982, Spain announced that it had selected the Hornet to replace the F-4Cs, F-5s, and Mirages operated by the *Eército del Aire Español*. Plans for 72 single-seaters and 12 two-seaters proved more than the Spanish economy could afford, and the 31 May 1983 order was for 60 single-seaters and 12 two-seaters. Spain also purchased 20 AGM-84 Harpoon antishipping missiles and 80 AGM-88 HARM antiradiation missiles to arm the EF-18s. The aircraft are capable of delivering laser-guided munitions and Mavericks, and the initial order included AAS-38 FLIR pods.

The Spanish EF-18s have been regular participants in United Nations (UN) and NATO operations over Bosnia and Yugoslavia. This EF-18A (C.15-13) shows the early standard markings for the type. The roundel and fin flash are high-visibility, although "low-vis" markings are sometimes used in combat zones. *(Boeing)*

Construcciones Aeronauticas SA (CASA) at Getafe is responsible for the maintenance of the *Eército del Aire* Hornets. As part of the offset agreement reached with Spain, CASA was awarded a contract to provide depot-level maintenance for U.S. Hornets serving with the U.S. Sixth Fleet in the Mediterranean. CASA also performed major overhauls of Canadian CF-18s while they were based in Germany. Somewhat later CASA was awarded a contract to manufacture the stabilators, flaps, leading-edge extensions, speedbrakes, rudders, and engine bay doors for all F/A-18s produced after 1985.[123]

The Spanish Hornets are generally referred to as EF-18A and EF-18B, the "E" standing for *España* (Spain). This does not follow the standard U.S. policy where "E" would usually designate an "electronic" variant of the aircraft (as in EA-6 or EF-111). Officially the Hornet is designated C.15 (single-seater) and CE.15 (two-seater), and assigned serial numbers C.15-01 through C.15-71 and CE.15-01 through CE.15-12.[124]

The first EF-18B (CE.15-01) was presented in a formal ceremony at St. Louis on 22 November 1985, and made its maiden flight on 4 December. The first few two-seaters were sent to Whiteman AFB, Missouri, where McDonnell Douglas trained the initial cadre of Spanish instructors. The first two-seater was flown to Spain on 10 July 1986. By early 1987, all 12 two-seaters had been delivered to Spain, and all single-seaters were delivered by the end of July 1990.

The Hornet serves with *Escuadron* (Squadron) 121 and *Escuadron* 122 of *Ala de Caza* (Fighter Wing) 12 at Torrejon de Ardoz AB, and with *Escuadron* 151 and *Escuadron* 152 of *Ala de Caza* 15 at Zaragoza-Valenzuela AB.

Escuadron 121 is dedicated to tactical support for maritime operations using AGM-84 Harpoon antiship missiles and laser-guided bombs. *Escuadron* 151 is primarily an all-weather fighter interceptor unit, and is also assigned to NATO's quick reaction alert force. *Escuadron* 152 is a SEAD (suppression of enemy aircraft defenses) unit and its Hornets are equipped with HARM antiradiation missiles. To date three of the Spanish Hornets have been lost in accidents.

An EF-18A+ (C.15-68) taxis out to conduct some air combat maneuvering training. Note the ACMI pod on the right wingtip launch rail, while the left rail carries a Sidewinder training round with a live seeker. Most of the Spanish aircraft now carry subdued squadron insignia on the vertical stabilizers. *(Neil Dunridge)*

The EF-18s are equipped with ALQ-126B jammers, and the last 36 aircraft also received ALQ-162(V) jammers. The earlier aircraft were later refitted with ALQ-162(V) systems to supplement their ALQ-126Bs. All 72 aircraft are equipped with the AN/ALR-67 RWR and ALE-39 dispensers. In 1993, plans were announced to upgrade the EF-18s to F/A-18C/D standards. McDonnell Douglas reworked 46 of the aircraft, while CASA modified the remainder. Changes involved adding new XN-6 mission computers and software improvements, an upgraded data bus, increased data-storage set, and wiring changes to the wing pylons to carry AIM-120 missiles. The NITE Hawk pods were upgraded to the AAS-38A standard, allowing the Spanish aircraft to laser-designate their own targets for the first time. Following the rework, the aircraft were redesignated EF-18A+ and EF-18B+.[125]

Spanish Hornets had been prepared to deploy to the Middle East in support of Operation Desert Storm, but were not needed. Instead the EF-18s deployed to Son San Juan AB, Majorca, to provide air cover for B-52s staging through Morón. The aircraft saw their first combat as part of Operation Deny Flight (*Operación Icaro* to the Spanish). On 1 December 1994 eight Spanish Hornets from *Escuadrons* 151 and 152 arrived at Aviano, Italy, to replace a squadron of U.S. F-15Es. The Spanish crews had just participated in Red Flag where they had extensively practiced SEAD tactics. The aircraft primarily flew CAP missions, but generally carried two 1,000-pound GBU-16 laser-guided bombs along with the normal AIM-7 and AIM-9s. The CAPs routinely lasted 4 to 5 hours with three in-flight refuelings. On 25 May 1995 two EF-18A+ aircraft dropped GBU-16s on Serbian-held targets near Pale, as part of a formation that included six U.S. F-16s. The Hornets self-designated the targets using their AAS-38A pods, while a second pair of Spanish Hornets provided SEAD coverage for the raid using HARMs. The following day the target was reattacked by another pair of EF-18A+ aircraft. Spanish Hornets also participated in Operation Deliberate Force in November 1995. Again, Spanish Hornets armed with HARMs provided SEAD coverage for forces attacking Serbian air defenses in Bosnia.

Worried because of delays in the Eurofighter 2000 program, Spain went in search of additional fighter aircraft. The U.S. Air Force offered 50 surplus F-16A/Bs, and the U.S. Navy offered 30 surplus F/A-18As. In September 1995, the *Eército del Aire* agreed to purchase 24 F/A-18As from the U.S. inventory. These aircraft were delivered to Spain beginning in December 1995, and equip *Escuadron* 211 of *Grupo* 21 based at Morón AB. *Escuadron* 211 functions primarily as an operational conversion unit. The last aircraft were ferried to Spain in December 1999. Before delivery the aircraft were upgraded at NAS North Island to the same EF-18A+ configuration as the rest of the Spanish Hornets. The ex-U.S. Navy aircraft are designated C.15A and carry serial numbers C.15A-73 through C.15A-96.

Kuwait

Kuwaiti interest in procuring a new fighter was prompted in 1987 by a perceived threat from Iran, largely because Kuwait had supported Iraq during the long Iran-Iraq war. Three aircraft types were evaluated to replace the existing Mirage F.1CKs and A-4KUs—the Mirage 2000, Tornado F.3, and F/A-18C/D.

In August 1988 the *Al Quwwat al Jawwiya al Kuwaitiya* (Kuwait Air Force) ordered 40 F/A-18s in a contract worth $489 million. Also included in the overall $1,900 million purchase were 120 AIM-9Ls, 200 AIM-7Fs, 344 AGM-65Gs, and 40 AGM-84Ds, as well as training and support.[126] The 32 single-seaters and 8 two-seaters were built to C/D standards, but were the first aircraft to be equipped with the more powerful F404-GE-402 engines and were also the first to introduce the Hazeltine APX-111 combined interrogator transponder (CIT). The aircraft include APG-65 radars, ALQ-165 ASPJs, ALR-67 RWRs, and

The first Kuwaiti F/A-18D (441), seen on 6 November 1993 at China Lake. The first two aircraft spent time at China Lake in weapons trials and electronic systems testing. Barely visible ahead of the windscreen are the five low-but-long blade antennas for the APX-111 combined interrogator transponder. The aircraft is also equipped with ALQ-165, evidenced by the antenna blister above the formation light strip on the nose. *(Craig Kaston via the Mick Roth Collection)*

ALE-47 dispensers. Internally, Boeing calls the aircraft KAF-18C/Ds. The aircraft are assigned Kuwaiti serial numbers 401 through 432 (single-seaters) and 441 through 448 (two-seaters).

The first Kuwaiti F/A-18D was scheduled to be delivered on 1 October 1991 to begin crew training in the United States, and the first six aircraft were due to arrive in Kuwait in January 1992. The Iraqi invasion of Kuwait on 2 August 1990 disrupted this schedule. All delivery (and payment) schedules were rearranged to allow Kuwait to cover wartime priorities.

The Kuwaiti Hornets (shown here at China Lake) were the most capable F/A-18s in existence at the time of their manufacture. They were the first production aircraft equipped with the -402 EPE engines, and included the most modern ECM suite available. Kuwait and Malaysia are the only two operators to paint their aircraft in something besides some variation of light gray. The Kuwait paint scheme is a mixture of blue and gray, unusual colors for a desert country. *(Craig Kaston via the Mick Roth Collection)*

The markings currently flown on the Kuwaiti Hornets differ somewhat from those applied to the first few aircraft at the factory. The large Kuwaiti flag has been made smaller and moved to the top of the leading edge of the vertical stabilizer. The Kuwaiti crest has been moved upward, and the serial number is now on the vertical instead of the rear fuselage. This quartet was photographed over the Persian Gulf. *(Peter Steinemann/Skyline APA)*

When the fighting was over, the two main air bases at Ali al Salem and Ahmed al Jaber had been largely destroyed by coalition air raids. With substantial assistance from the U.S. Army Corps of Engineers, the bases were rebuilt and prepared for the arrival of the F/A-18s.

At the end of Operation Desert Storm, the production program went forward. The first KAF-18D (441) made its maiden flight on 19 September 1991, and was formally presented to the *Al Quwwat al Jawwiya al Kuwaitiya* on 8 October. The first two D models would spend time at China Lake over the next 2 years in various weapons and ECM tests. The first three Hornets were flown to Kuwait on 25 January 1992, and the last arrived in Kuwait on 21 August 1993. An option for 32 additional Hornets lapsed in 1992 without being exercised.

The multiple ejector rack on the inboard pylons signals this Kuwaiti F/A18C (412) has been engaging in bombing practice. The -402 EPE engines provided a worthwhile improvement in hot weather performance over the -400 used in American Hornets at the time. *(Peter Steinemann/Skyline APA)*

A Kuwaiti F/A-18C (418) shows the forward-fuselage searchlight that is also used by Canada, Finland, and Switzerland. Note the slight deflection of the leading- and trailing-edge flaps while this Hornet climbs vertically. *(Peter Steinemann/Skyline APA)*

The first Hornets were assigned to No. 25 Squadron operating from Kuwait International Airport, although the squadron subsequently moved back to the Ali al Salem military air base after it was rebuilt. Later aircraft went to No. 9 Squadron which also initially operated from Kuwait International, but moved to Ahmed al Jaber AB when it was reopened on 7 February 1994. Both of these squadrons previously operated A-4KU Skyhawks, which were sold to the Brazilian Navy in 1998.[127]

The Kuwaiti Hornets began flying missions in support of Operation Southern Watch over Iraq in late 1993. Since then they have engaged in regular exercises with other Gulf air forces, including deployments to Bahrain.

Switzerland

To fulfill a requirement for a *Neue Jagdfugeuge/Novel Avion de Combat* (New Combat Fighter), the Swiss government evaluated the Dassault Mirage 2000, Israel Aircraft Industries Lavi, Northrop F-20, and Saab JAS-39 Gripen. Surprisingly, these aircraft were all deemed unsatisfactory by the Swiss. During April and May 1988, the Swiss government held a fly-off between the F-16 and F/A-18. On 3 October 1988 the government announced that 26 F-18Cs and 8 F-18Ds would be procured for the *Schweizerische Flugwaffe/Troupe d'Aviation Suisse* (Swiss Air Force). The Swiss Hornets are designated F-18, rather than F/A-18, because of the Swiss emphasis on air defense rather than on strike warfare.[128] Internally, Boeing calls the aircraft SF-18C/Ds. The aircraft were assigned Swiss serial numbers J-5001 through J-5026 (single-seaters) and J-5231 through J-5238 (two-seaters).

In 1991 the competition was reopened so that the MiG-29 and the Dassault Mirage 2000-5 could be considered. However, even a personal appeal on the part of French President Francois Mitterand could not overturn the original plan to buy the Hornets. The formal contract was originally expected to be signed in 1992; however, the Hornet order remained extremely controversial. The procurement was the subject of a popular referendum held on 6 June 1993, which finally approved the program.

The SF-18D (J-5231) made its maiden flight on 20 January 1996, with a ceremonial roll-out in St. Louis on 25 January 1996. The first SF-18C (J-5001) made its maiden flight on 8 April 1996. Both aircraft were initially retained by McDonnell Douglas for weapons systems testing, finally being delivered to Switzerland on 16 December 1996. The remaining 32 F-18s were assembled at the Swiss Aircraft and Systems Co. (SASC) in Emmen, largely from components supplied by McDonnell Douglas. Over 60 Swiss companies also provided parts such as wing components, tail surfaces, and landing gear. The engines and avionics

The Swiss F-18s are very similar to the Kuwaiti aircraft, with -402 EPE engines, ALQ-165, and the searchlight on the forward fuselage. Switzerland uses an overall light gray paint instead of the U.S. two-tone gray, presenting a much cleaner appearance. The national insignias are in full color. Note the inboard deflection on both rudders while the aircraft is taxiing. *(Neil Dunridge)*

Swiss aircraft are designated F-18 rather than F/A-18 to emphasize their defensive roles, although they are fully equipped to conduct offensive operations. The only truly modern piece of equipment the Swiss aircraft appear to be missing is the APX-111 CIT, evidenced by the lack of the five "bird slicer" antennas on the top of the nose. *(Christoph Kugler)*

were excluded from this offset agreement, however, as the Swiss order was too small to warrant manufacturing rights for such complex components.[129]

The first SF-18D (J-5232) assembled in Switzerland made its maiden flight 3 October 1996, and the aircraft was officially turned over to Corps Commander Fernand Carrel, the senior Swiss Air Force officer, at Lucerne on 23 January 1997. Three squadrons at Payerne, Sion, and Meiringen operate the Hornet in the air defense role. The total cost of the program, including aircraft, spares, support, training, and infrastructure modifications, is $2,300 million.

Swiss aircraft differ from the standard C/D by being designed to a +9g load limit instead of the +7.5g limit for all earlier aircraft. These represent the first Hornets stressed to withstand the higher limit.[130] They are expected to maintain a 30-year, 5,000-flight-hour ser-

Like many modern combat aircraft, the color of the radome on this Swiss F-18C (J-5021) does not quite match the rest of the aircraft. This is because the radomes are usually molded in color since they cannot be painted with normal paint because of their dielectric nature. *(Christoph Kugler)*

This Swiss F-18C (J-5009) shows that the Swiss did not delete the catapult launch bar on the nose landing gear, perhaps fearing the same shimmy that afflicted the Australian aircraft. Except for having some components of local manufacture, the Swiss aircraft are similar to late production U.S. C/D models. *(Christoph Kugler)*

vice life without additional modification. The Swiss Hornets also have an air-intercept spotlight fitted to the left side of the nose. The Swiss specified the APG-73 radar, ALQ-165 ASPJ, ALR-67 RWR, and ALE-47 dispenser for the aircraft. The aircraft are powered by F404-GE-402 EPE engines and are equipped to launch AIM-120 AMRAAMs. Although the Swiss use the aircraft primarily in a defensive role, they are fully capable of carrying the AAS-38A LDT/R and AAR-50 thermal imaging navigation set (TINS) pods.[131]

Although seldom seen with anything other than Sidewinders on the wingtips, the Swiss F-18s (J-5233 seen here) are fully capable of carrying AIM-7 and AIM-120s as well. In fact, despite their mainly defensive role, the Swiss aircraft can carry the whole range of offensive weapons used by U.S. Hornets, and the Swiss procured some number of AAS-38B and AAR-50 pods. *(Christoph Kugler)*

The last two Swiss F-18s that had been conducting flight testing at Pax River and China Lake returned to Switzerland in March 1998, completing an active 2-year test program. Weapons separation testing and missile launchings were conducted, with emphasis on the AIM-9 and AIM-120. The aircraft went through a series of radar cross-section measurement flights and also used an anechoic chamber to evaluate their electronic warfare systems.[132]

Korea

The F/A-18C/D was announced as the winner of the Korean Fighter Program in December 1989. The Hornet had experienced stiff competition from the F-16, a type that was already in service with the Republic of Korea Air Force (RoKAF). The South Korean government regarded the all-weather performance of the F/A-18 as superior to that of the F-16, which meant that it would be better equipped to carry out missions during the bad weather typically experienced during the winter months. Also, since the F/A-18 was able to carry a FLIR pod but the F-16[133] could not, the Hornet could be more effective than the F-16 against North Korea's fleet of Antonov An-2 fabric-covered biplanes, which have a very small radar cross-section but which can be spotted via infrared. Finally, the Koreans felt that the F/A-18 would be more capable than the F-16 against North Korean MiG-29s in air-to-air combat.

A total of 120 F/A-18s were ordered. The first 12 were to be manufactured by McDonnell Douglas, with the next 36 being supplied in kit form for assembly by Samsung Aerospace Industries at Sachon. The final 72 were to be manufactured from scratch under license by Samsung. General Electric would provide 27 F404-GE-402 engines, with Samsung building 10 engines from General Electric–supplied kits, and 144 being wholly manufactured in Korea.

However, the costs to establish the required manufacturing capabilities escalated rapidly, and by the end of 1991 the F/A-18s were 50 percent more expensive than when initially ordered. Additionally, there were bribery and corruption charges being levied against various high-level former government officials. In March 1991 the South Korean government canceled the procurement and announced it would purchase additional F-16C/Ds instead.

Finland

In early 1989 the *Ilmavoimat* (Finnish Air Force) began to look for a replacement for its MiG-21s and Saab J-35 Drakens. On 23 February 1990 Finland released requests for quotations (RFQ) for 20 single-seaters and 5 two-seaters to Dassault Aviation, General Dynamics, and Saab-Scania. Although not specifically named in the RFQ, the choices were the Mirage 2000-5, F-16C/D, and JAS-39 Gripen. During late 1991 and early 1992, the *Ilmavoimat* conducted flight evaluations of the three aircraft in two phases: in the country of manufacture first, then in Finland. In addition, the MiG-29 was evaluated in Russia.[134]

Quotations were received on 31 October 1990, but none was considered totally satisfactory. An alternate RFQ for 60 single- and 7 two-seaters was released to the same manufacturers on 31 January 1991. This second request was also released to McDonnell Douglas on 12 April 1991.

On 12 February 1992 a Marine F/A-18D (BuNo 164652) landed at Halli and participated in 15 evaluation flights over the next 3 weeks. On 6 May 1992 the Finnish Prime Minister, Esko Aho, announced Finland's intention to acquire 57 F-18Cs and 7 F-18Ds. The Finnish aircraft were intended strictly for defensive purposes, and are designated F-18 instead of F/A-18. Finland did not order AAS-38 or AAR-50 pods, reflecting the defensive nature of the aircraft (it should be noted the aircraft are wired to accept the pods if the need ever arises). Internally, Boeing calls the aircraft FN-18C/Ds. The procurement included 3 years of spare

Another country that chose to designate its Hornets F-18 instead of F/A-18 is Finland. Unlike the Swiss, however, Finland did not purchase sensor pods. Nevertheless, the Finnish F-18s (HN-465 here) are very capable aircraft, complete with -402 EPE engines and ALQ-165 ASPJ. Like the Swiss aircraft, all Finnish Hornets are painted an overall light gray, with small full-color national markings. *(Finnish Air Force)*

parts, depot and organizational logistics for 3 years, product support through delivery, training of aircrews as well as maintenance and assembly personnel, pilot equipment (helmets etc.), and 300,000 pages of documentation. Most of the training and documentation is in English.[135]

The decision to purchase the F-18 came as something of a surprise, since the original intent was to choose an inexpensive and lightweight fighter-interceptor. The F-18 is neither, being relatively heavy and clearly a more expensive fighter than the competition. The Finnish Air Force subsequently listed the decisive factor for selection of the F-18 as "the superior quality/cost ratio, including both purchase costs and life-cycle costs." According to Major-General Heikki Nikunen, the maintenance man-hour per flight hour ratio was the lowest of the aircraft evaluated. This relatively low maintenance requirement makes the use of reserve and conscript maintenance personnel a realistic possibility.

A letter of acceptance was signed on 5 June for a total of 64 aircraft, with the first 7 F-18Ds (HN-461 through HN-467) being built by McDonnell Douglas in St. Louis, and the remaining 57 F-18Cs (HN-401 through HN-457) all being assembled by Valmet Aircraft Industries (now Finavitec) of Kuorevesi from McDonnell-supplied kits. The F404-GE-402 engines would also be assembled locally, with General Electric supplying 137 kits. The AIM-120s and AIM-9 missiles that arm the Finnish Hornets are manufactured in the United States.[136]

The *Ilmavoimat* specified the APG-73 radar, which, at the time, had not yet been installed on U.S. aircraft, and the searchlight on the left side of the nose. The Finnish F-18s are equipped with the ALQ-165 advanced self-protection jammer (ASPJ) under a $128 million contract signed on 30 September 1994. The *Ilmavoimat* was the first production customer for the ALQ-165, which had been canceled by the United States in 1992. Other ECM equipment includes the ALR-67 RWR and ALE-47 dispensers, and the aircraft are also equipped with a Finnish-built data link.

Three *Ilmavoimat* fighter squadrons operate the Hornet: 2. *Lentue/Hävittäjälentolaivue* 11 (2 Flight of Fighter Squadron 11, usually abbreviated *HävLLv*) at Rovaniemi, 2. *Lentue/Hävittäjälentolaivue* 21 at Tampere-Pirkkala, and 2. *Lentue/Hävittäjälentolaivue* 31

The first Finnish F-18C (HN-401). All of the Finnish single-seaters were assembled in Finland, including the engines and radar units. The seven two-seaters were all assembled by McDonnell Douglas in St. Louis. The Finnish Air Force elected not to procure an optional "cold weather" package developed by McDonnell Douglas, but all reports are that the aircraft are working well in the severe climate. *(Finnish Air Force)*

at Kuopio-Rissala. The first two previously flew the J-35 Draken, while the third operated the MiG-21bis; they are the only fighter squadrons operated by the Finnish Defense Forces.[137] The *Ilmavoimien Koelentokeskus* (Air Force Flight Test Center) at Halli also operates a few Hornets.[138]

The first FN-18D (HN-461) made its maiden flight in St. Louis on 21 April 1995, piloted by Fred Madenwald and Lieutenant Commander Dave Stuart. Afterward, it received its gray paint and Finnish military insignia, and its official rollout was on 7 August. By the end of the month, five of the seven D models had been flown.

The first four FN-18Ds (HN-462, HN-464, HN-465, and HN-466) were delivered on 7 November 1995 to the Satakunta Air Command's Pirkkala Air Base south of Tampere. Over 5,000 observers were on hand to watch the aircraft land after their 9-hour, 35-minute, 5,100-mile flight across the Atlantic. The aircraft were flown by American pilots from VFA-125 "Rough Raiders" with Finnish pilots in the rear seats. Each aircraft carried two 330-gallon external tanks, and had been refueled nine times by a KC-10 tanker during the flight. The KC-10 was itself refueled twice by a KC-135. Each fighter burned about 7,920 gallons of fuel on its way to Finland, averaging 14 gallons per minute.

The first Finavitec-assembled FN-18C (HN-401) was delivered 3 months ahead of schedule on 28 June 1996, and it entered active service the same week. The delivery ceremony at the Air Force's Kuorevesi plant in Halli was attended by Finnish President Martti Ahtisaari and Minister of Defense Anneli Taina. Seven aircraft entered *Ilmavoimat* service in 1995, 4 in 1996, 10 in 1997, 13 in 1998, and 18 in 1999, and 12 are scheduled for 2000.

Malaysia

After months of debate, on 29 June 1993 the Defense Minister announced that Malaysia would order both MiG-29s and F/A-18s for the *Tentara Udara Diraja Malaysia* (TUDM, or Royal Malaysian Air Force). On 9 December 1993 Malaysia signed a letter of offer and acceptance for eight F/A-18D Hornets, although the order was not actually placed until 7 April

The eight Malaysian F/A-18Ds are all assigned to the No. 18 Night Strike Fighter Squadron at But-terworth, near Penang. These aircraft are all painted an overall gunship gray, similar to the one used on the F-15E Strike Eagle. This F/A-18D (M45-01) shows the standard markings, with a small full-color national insignia on the forward fuselage, a color fin flash, and black serial numbers on the vertical stabilizer. Note the deployed speed brake between the vertical stabilizers. *(Peter Steinemann/Skyline APA)*

1994. Internally, Boeing calls the aircraft MAF-18Ds. The total cost of the F/A-18 order was $600 million, with a $250 million offset deal. The contract included a substantial number of AIM-7, AIM-9, AGM-65, and AGM-84 missiles. On 7 June 1994 Malaysia ordered 18 MiG-29R Fulcrums from Russia for $550 million.[139] The MiGs are used primarily in the interceptor role, while the Hornets are used for interdiction, night attack, and maritime strike.

The Malaysian aircraft are generally similar to the Kuwaiti standard but without the nose-mounted searchlight. The aircraft include -402 EPE engines and ALQ-165 ASPJ. *(Peter Steinemann/Skyline APA)*

The first MAF-18D (M45-01) was rolled out in a ceremony in St. Louis on 19 March 1997, having made its maiden flight on 1 February 1997. The aircraft had spent most of the intervening period conducting electronic warfare tests at China Lake. The first four Hornets (M45-01 through M45-04) were delivered to No. 18 Night Strike Fighter Squadron at Butterworth, near Penang, on 19 March 1997, with the last four aircraft (M45-05 through M45-08) arriving on 28 August. The unit was declared operational on 1 March 1998.

The Malaysian Hornets are equipped with the AGP-73 radar, ALQ-165 ASPJ, ALR-67 RWR, ALE-47 dispensers, and F404-GE-402 engines. Malaysia has expressed interest in obtaining 10 to 16 additional Hornets, but the Asian financial crisis has forced a postponement of a decision.

Thailand

In May 1996 Thailand signed a letter of offer and acceptance for four F/A-18Cs and four F/A-18Ds for delivery between February and October 1999. The Hornets were to include the APG-73 radar, ALQ-165 ASPJ, ALR-67 RWR, ALE-47 dispensers, specialized data links, additional radios, and F404-GE-402 engines.

The 1998–1999 financial crisis in Asia forced Thailand to slash its $4 billion defense budget by nearly one-fourth, resulting in the F/A-18 order being canceled. The U.S. government agreed on 13 March 1998 to relieve Thailand of the approximately $250 million in remaining financial liability for the aircraft. Thailand was not reimbursed for $70 to 75 million already paid, although this has been reduced by other countries purchasing some of the spares, etc. The value of the Thai contract was $578 million according to Boeing, including eight aircraft, four spare engines, and five Harpoon missiles.[140]

The Pentagon agreed to purchase the partially completed aircraft, finishing them to standard U.S. F/A-18D configuration and using them as attrition reserves for the Marine Corps. The aircraft will be delivered in early 2000.[141]

Singapore

McDonnell Douglas was less successful in Singapore. Already an F-16A/B operator, Singapore evaluated the F/A-18 and MiG-29 before ordering F-16C Block 50s in July 1994. One of the major factors in the decision was a greater than 100 percent offset deal by Lockheed Martin where goods and services worth more than the purchase price of the aircraft would be bought in Singapore.

Chile

An initial *Fuerza Aerea de Chile* (Chilean Air Force) comparison of candidates in its fighter competition has rated the F/A-18C first, JAS 39 Gripen second, F-16C third, and Mirage 2000-5 last, according to U.S. officials.[142]

The Chilean Air Force's technical evaluation, which established the ranking, is a major step in selecting a new fighter, according to military and aerospace officials. The initial contract would be for 16 to 20 fighters, but this is seen as the first step in larger Chilean and regional sales. Boeing offered the "Thai configuration" F/A-18 with specialized data links, optional radios, ALQ-165 ASPJ, and -402 engines.

The fly-away, per unit cost for the aircraft is thought to be roughly $27 million for the F-16, $33 million for the Gripen, $35 million for the F/A-18, and more than $40 million for the Mirage. Contractors also said the purchase price of the aircraft could change, but cuts would probably be reflected in a higher price for the spares and support package.[143]

This AF-18A (A21-30) from No. 75 Squadron flies over the Olgas *(Kata Tutja* in Aboriginal). Note the lack of national insignia on the upper wing surfaces. *(RAAF/LAC Peter Gammie via Chris Hake)*

A Finnish F-18C takes off during winter. Note the exhaust heat from the engines and the wingtip vortices. Snow can be seen covering much of the ground around the air base. The Finnish Hornets would operate from well-prepared road bases during time of crisis. *(Finnish Air Force)*

The annual NATO "Tiger Meet" often produces some interesting paint schemes among the participants, and this Canadian CF-18A (188769) shows the extent some crews will go to. Note the tiger head on the inside of the vertical stabilizer. This aircraft was photographed in August 1991 at RAF Fairford. *(Pat Martin via the Mick Roth Collection)*

Poland

Cash-strapped Poland has expressed an interest in purchasing up to 60 new fighters to replace its current force of Russian aircraft. In mid-1999 the United States offered to lease seven surplus F/A-18A/Bs to Poland as a possible prelude to purchasing new F/A-18C/Ds in the future. Alternatively, the United States also offered the same deal using F-16s. No immediate decision is expected to be forthcoming from Poland until its economy recovers somewhat.[144]

The Kuwaiti Hornet order was postponed by the Gulf War of 1991, which resulted in the destruction of the two air bases the aircraft were to be assigned to. When the F/A-18s began to arrive they were temporarily assigned to the Kuwait International Airport, which had been reopened shortly after the end of the war. Eventually the two military air bases were rebuilt and the aircraft transferred to them. *(Peter Steinemann/Skyline APA)*.

Super Hornet: The F/A-18E/F

During the 1980s, McDonnell Douglas marketed an advanced version of the F/A-18 known as the *Hornet 2000.* The aircraft used a larger wing and stabilator, two fuselage plugs to carry additional internal fuel, more powerful engines, and an improved cockpit. McDonnell Douglas submitted the concept to the Royal Air Force and Luftwaffe as an alternative to the European Fighter Aircraft (EFA—now the Eurofighter 2000) without success. During early 1987 a Pentagon delegation went to Europe in an unsuccessful attempt to interest the French in a Hornet 2000 codevelopment program. Later in 1987, a study conducted by the Naval Air Systems Command, McDonnell Douglas, and Northrop focused on potential Hornet upgrades and derivatives.

On 7 January 1991 the troubled General Dynamics/McDonnell Douglas A-12 Avenger II stealth attack aircraft program was canceled. The aircraft had been intended to replace the A-6 but was clouded by government accusations of gross mismanagement by the contractors. The contractors disagreed with this assessment and took the government to court over the cancellation, subsequently winning a favorable settlement involving several billion dollars.

Since an A-6 replacement was still needed, a new program was undertaken. Originally known as the A-X (Attack, Experimental), the program was later redesignated A/F-X (Attack/Fighter, Experimental) when it was merged with the Naval Advanced Tactical Fighter (NATF), which was intended to produce an F-14 replacement. At this time, the leading contender was the Lockheed/Boeing AFX-653, which was essentially a navalized two-seat F-22. Like the F-22 program itself, the A/F-X costs escalated quickly, unfortunately at a time when the real-value defense budget was shrinking.

When the A/F-X was canceled in late 1991, McDonnell Douglas proposed an alternative—a stretched version of the F/A-18 optimized for the attack role, much the same as the derivatives studied in 1987. The internal improvements made when the C/D was developed largely satisfied the Navy from a systems perspective, being well integrated and easily upgradable. However, the C/D still suffered from the airframe limitations of the A/B—limited bring-back capability and still a little short on range. The McDonnell Douglas

Even though the F/A-18E/F was a "derivative" aircraft, the E/F models spent a lot of time in various wind tunnels. Here a high-fidelity model is prepared for measuring the drag associated with various weapons loads. Note the AIM-120s on the fuselage stations, with AIM-7s on the outer wing stations. *(NASA via the Tony Landis Collection)*

proposal involved wrapping a new airframe around the proven C/D systems and new engines.

The Navy was impressed by the proposal and believed they could muster political support for a "derivative" aircraft that represented a low-risk development effort. On 12 May 1992 the Navy announced their intention to procure the F/A-18E/F. The total program cost was estimated to be $63,090 million (FY96 dollars) consisting of $5,783 million in development costs, and $57,310 million in procurement costs for 1,000 aircraft.[145] The initial operational capability was scheduled for 2000, with the first operational carrier-based squadron deploying in 2003. Procurement of the F/A-18E/F was expected to continue through 2015.[146]

One of the more important goals of the project was a 40 percent increase in range over the F/A-18C, finally curing the Hornet's only major shortcoming. In 1992 the F/A-18's operational requirements specified a combat radius of 410 nautical miles for fighter missions and 430 nautical miles for attack missions. These requirements were never achieved by the F/A-18C/D, whose range/payload capabilities have been reduced by weight growth due to equipment added in successive upgrades since 1982, when its combat radius was 366 nautical miles for fighter missions and 415 nautical miles in attack missions. In 1992 the Navy projected the F/A-18E/F's fighter combat radius to be about 420 nautical miles, with an attack radius of about 490 nautical miles—exceeding the original F/A-18C/D requirements for these missions.

Although the Navy sold[147] the program as a "derivative," the F/A-18E/F is essentially a new aircraft, and bears little more than a passing aerodynamic resemblance to previous Hornets.[148] However, most of the initial avionics suite planned for the E/F is common with late-model C/Ds, and avionics and weapons system development is generally the most expensive (and risky) part of a new aircraft. In April 1992, the Defense Acquisitions Board

This F/A-18E model is undergoing inlet testing at the Air Force's Arnold Engineering Development Center near Tullahoma, Tennessee. The new intakes of the E/F allow much greater airflow to the engines, a necessity given the 5,000-lbf increase in thrust of the new F414s. *(AEDC via the Tony Landis Collection)*

(DAB) approved entering the engineering and manufacturing development (EMD) phase. A $4,880 million EMD contract was signed on 7 December 1992, calling for 3 non-flyable test articles, 5 single-seat F/A-18Es, and 2 two-seat F/A-18Fs. At the same time, General Electric was awarded a $754 million contract to develop the new F414 engine.

The critical design review was held 13 to 17 June 1994, with the F/A-18E/F passing all schedule, cost, technical, reliability, and maintainability requirements. Production of the first center/aft fuselage by Northrop Grumman began on 24 May 1994, and McDonnell Douglas opened the Super Hornet assembly line in St. Louis on 23 September 1994.[149] On 18 September 1995 the first F/A-18E was rolled out in a ceremony in St. Louis. Secretary of the Navy John H. Dalton said, "The development process for the F/A-18E/F is one of the first, and by far the foremost example of our success in Acquisition Reform in the Navy." Dalton called the aircraft a "remarkable achievement," adding that "in all respects this aircraft has been on target: It is on schedule, on budget, and underweight."[150]

The Improved C/D Debate

In March 1996 a 3-year flight test program began at Pax River. Ironically, at the same time the General Accounting Office (GAO) suggested that the Super Hornet be canceled and that F/A-18C/D purchases continue until the Joint Strike Fighter (JSF) became available some time between 2008 and 2010.[151]

In the GAO's estimation, with "minor upgrades" and "minimal changes," the F/A-18C/D could carry more than 10,000 pounds of ordnance with stronger landing gear and larger 480-gallon external fuel tanks. To the GAO, the Super Hornet's estimated $11.1 million per aircraft premium over the projected $32.5 million for the Navy's JSF was at the crux of the discussion. The GAO clearly considered the Super Hornet only a "marginal improvement" over its F/A-18C/D predecessors.

The Navy disagreed. Expecting minor upgrades to perform miracles on an aging aircraft design was a futile effort. As the Navy saw it, the GAO's recommendation of increasing landing weight to boost recovery payload was simply untenable considering the F/A-18C/D design limitations and the margins required by existing safety regulations. Engineering evaluations showed no economical modifications that would significantly enhance the F/A-18C/D.

However, largely in response to these evaluations, the Navy introduced a 990-pound increase in the F/A-18C's landing weight during operations over Bosnia, primarily to allow recovery of unexpended munitions. Prior to this, a late-production F/A-18C could not return to its carrier at night with an AGM-84E SLAM or AGM-88 HARM missile.[152] The increase was accomplished with minor flight control software and procedural changes. The Navy felt that beyond this, nothing more could realistically be done. Performance limits were stretched as far as they would go with the current F/A-18, and any further recovery payload increases would require significantly strengthening the internal structure to accommodate them.

Engineers were well aware of the spiral that would follow. Reinforcing the airframe inevitably means adding weight. Added weight raises the approach speed. Compensating with a corresponding increase in wing area increases drag. By the time all the emerging flight dynamic challenges were worked out, an altogether new aircraft was taking shape. This was basically how the F/A-18E/F configuration had been arrived at in the first place.

There has been considerable debate over the F/A-18E/F's ability to replace the F-14 fleet defense fighter. Proponents of the Super Hornet acknowledge that the aircraft will lack

This EMD F/A-18E shows an impressive load-out of air-to-air missiles. Visible in this photo are an AIM-9 on the wingtip, an AIM-7 on the outboard pylon, and two AIM-120Cs on each of the middle and inner pylons. The wingtip formation light strip is mounted on small vertical plates above and below the wing just inboard of the missile rail. The prominent snag on the wing leading edge shows up well in this photo. The light-colored wedge shape on the leading edge of the LEX is a dielectric panel covering various ECM antennas. *(Boeing)*

some of the F-14D's mission capabilities in range, payload, speed, and all-weather capability, but they argue that the demise of the Soviet threat has fundamentally changed the requirements that carrier-based fighters must meet. Currently armed with the AIM-54C Phoenix missile, the F-14 was designed in the early 1970s to defend U.S. battle groups in midocean waters against Soviet aircraft armed with long-range antiship cruise missiles. Some advocates of the F/A-18E/F argue that such a long-range, midocean air defense scenario is now implausible, as reflected in the Navy's current emphasis on littoral operations in third-world conflicts.[153]

Proponents of the F-14D emphasize the aircraft's inherent advantages in range/payload/endurance, which they believe may still be needed in some threat scenarios where sophisticated aircraft and air defense systems are likely to be available to hostile forces in regional conflicts involving the U.S. Navy. Tomcat supporters also argue that the long-range detection and multitarget tracking capabilities of the F-14D's APG-71 radar can be used to command and control other aircraft, thus extending fleet surveillance. Hornet supporters counter that the APG-73 is rapidly becoming almost as capable as the APG-71, although its limited antenna size will always restrict its range somewhat. They also argue that the AIM-120, and future longer-range variants, are generally more capable missiles than the Phoenix. In any case it is a moot point, since the F/A-18F is currently scheduled to replace all F-14s by 2005. The projected firepower from Super Hornets operating from aircraft carriers is a key contributor to the Navy's *Joint Vision 2010* concepts of dominant maneuver and precision engagement.[154]

F/A-18E/F

The F/A-18E/F is an upgraded and larger version of the F/A-18C/D, with increased range and payload capacity. The single-seat F/A-18E and two-seat F/A-18F will replace some F/A-18C squadrons and all F-14 squadrons in service with the U.S. Navy. Eventually, carrier air wings will have two 12-aircraft squadrons of F/A-18Cs (Lot XII and later), one squadron of 12 F/A-18Es, and one squadron of 14 F/A-18Fs.

Boeing and its industry partners have developed a remarkably improved aircraft, taking advantage of more than 3,000,000 flight hours experience with earlier F/A-18s.[155] In what might be a first for a modern combat aircraft, a Navy press release fairly shouts that the ". . . new Super Hornet is *under-weight, under-budget, and continues to meet all scheduling requirements on time.*" The emphasis is theirs.[156]

The F/A-18E/F program is a successful example of the defense reform initiatives (DRI) in action. In 1990, Congress established an engineering and manufacturing development cost ceiling of $4.88 billion, and a production cost no more than 125 percent of the C/D-model's recurring flyaway cost. This forced the Navy and Boeing to apply the acquisition reform principle of cost as an independent variable (CAIV) from the program's inception. This led to an affordable program that has completed 94 percent of its developmental tests while adhering to the original schedule laid out in 1992. At the end of 1999, the Super Hornet program was 12 percent below the congressional cost cap.[157]

The Super Hornet has a 2.83-foot fuselage plug in order to provide a 33 percent increase in internal fuel (from 10,381 pounds to 14,500). This will allow a fleet air defense F/A-18E/F carrying four AIM-120s, two AIM-9s, and external tanks to loiter on station for 71 minutes at a distance of 400 nautical miles from its carrier, compared to only 58 minutes for the F-14D. Unlike earlier Hornets, the E/F model equipped with an air-refueling system (ARS) pod on the centerline will be able to serve as a tanker to refuel other aircraft.

The wing is 25 percent larger, with an increase in span of 4.29 feet and an additional 100 square feet in area. The new wing has no twist or camber and has an outboard leading edge flap chord extension, leading to a definite snag, reminiscent of the one originally included

The rear fuselage of the sixth F/A-18E is lowered onto an assembly jig in St. Louis during June 1998. Note the attachment point for the horizontal stabilator. The dark skins on the vertical stabilizers are graphite-epoxy composite, while the lighter skin is aluminum alloy. *(Boeing)*

E6 (BuNo 165533) in the assembly jig, waiting for the final attachment of the rear fuselage. Note the electronic compartments stretching almost the entire length of the forward fuselage. All of these are at a convenient working height when the aircraft is on the ground. Unlike assembly areas of old, modern aircraft plants are nearly spotlessly clean. *(Boeing)*

The sixth F/A-18E (BuNo 165533) made its maiden flight without paint. E6 is the first production aircraft. Without its underwing pylons the F/A-18E is a very clean looking aircraft and belies its large size. In reality the aircraft is as large as an F-15. *(Boeing photo by Kevin Flynn)*

on the FSD F-18As. The area of the leading edge extensions (LEXs) was increased 34 percent, from 56.0 to 75.3 square feet, in order to ensure full maneuverability at 40° angle of attack.

While Boeing and the Navy consider the performance improvements to the Super Hornet to be significant, they did not lose sight of the fact that the AIM-9X and AIM-120 with off-boresight capability and helmet-mounted cueing make aircraft performance much less critical in the vast majority of combat scenarios. For that reason, the Super Hornet is stressed to +7.5g, rather than the +9g in most modern fighters, including some export versions of the F/A-18C/D. "You pay a lot for that extra few gs in weight and cost, when new missile technology negates the value," Rear Admiral John (Carlos) Johnson, director of the

E5, accompanied by a nearly hidden F1, banks over the desert near China Lake. The number of photoreference markings on almost every conceivable surface indicates the aircraft was involved in stores separation tests at the time, although only two camera locations are evident—one under each engine. The two ALE-47 dispensers under the air intake are visible. *(Boeing)*

Aviation Plans and Requirements Branch, said.[158] This is also largely the reason that other new technologies, such as thrust vectoring, were not incorporated into the Super Hornet.

The F/A-18E/F is equipped with an additional stores station under each wing, although it can carry only 1,150 pounds instead of 2,600 pounds. This raises the total external stores capability to 17,750 pounds. The number of ALE-47 countermeasures dispensers has been increased from two to four, increasing the load from 60 to 120 canisters.

Although empty weight has increased to 30,500 pounds, the aircraft is some 900 pounds under its specification weight. The maximum gross weight is increased by about 11,600 pounds. The Super Hornet's maximum carrier recovery payload of 9,317 pounds is 3,500 pounds greater than the F/A-18C/D. This translates to an almost 200 percent increase in "actual" payload (weapons, etc). The maximum carrier-landing weight is 42,900 pounds, while maximum takeoff weight is 66,000 pounds.

The F/A-18E/F is powered by a pair of 21,890-lbf General Electric F414-GE-400 turbofans. The F414 is a major derivative of the F404 used on earlier F/A-18s, and incorporates many of the features developed for the F412 engine planned for the canceled A-12. The engines are equipped with full-authority digital electronic control (FADEC), the first on an operational fighter. The FADEC has worked well in the most trying situations, including inverted spins above 40,000 feet and tail slides, and there are no throttle restrictions or limitations on the Super Hornet. The aircraft also uses a "throttle-by-wire" system with no mechanical linkage between the throttle and engine. A completely redesigned trapezoidal engine air intake replaces the D-shaped intakes of earlier Hornets, providing 18 percent more airflow to the engines, better performance at high speed, and a reduced radar cross section.[159]

The area of the vertical stabilizers has been increased by 15 percent, while the rudders have been increased in area by 54 percent and can be deflected up to 40°, a 10° increase. The stabilators are made of improved composites and their area is increased by 36 percent.

The most visible change to the Super Hornet is its new wedge-shaped air intakes, replacing the D-shaped ones used on all earlier models. These intakes can accommodate the higher airflow requirements of the F414 engines, and also offer a reduction in the aircraft's radar cross section (RCS) due to careful alignment of the various angles used in their design. The faceted design of the main landing gear doors can also be seen here, this further contributing to reducing the RCS. *(Mick Roth)*

NASA's HARV F/A-18A made 385 flights investigating flying qualities at extremely high angles of attack, including thrust vector control. This aircraft was not yet fitted with the LEX fence, and the smoke trail shown here clearly impinges on the vertical stabilizer. The vertical stabilizer mounts would be reinforced on all Hornets to withstand the stresses encountered during high-angle-of-attack flight. *(NASA)*

When the YF-17 was rolled out at Hawthorne, it was painted in an overall silver finish with dark blue trim, unlike the YF-16 that was revealed in a bright red, white, and blue paint scheme. Note the simple undercarriage on the YF-17 demonstrator. *(Peter Bergagnini via the Mick Roth Collection)*

This photo of a Blue Angel also shows the vortices coming off the LEX, this time in the form of condensation. The single trail coming from the rear of the fuselage is smoke used during the show, while the two wingtip trails are more condensation, a phenomenon often encountered in the moist air found at most naval air stations. *(Dennis R. Jenkins)*

The standard U.S. Navy paint scheme is the usual shades of gray. Each of the deck crewmember aboard a carrier wears a specific-color shirt to signify the member's role in servicing the aircraft. This F/A-18A from VFA-113 is getting ready to be catapulted off the USS *Constellation. (U.S. Navy/DVIC DN-SC-86-01301)*

The fourth EMD Super Hornet (BuNo 165168) was to investigate the high-angle-of-attack performance of the E/F. Note the attachment location for the spin recovery parachute on the upper aft fuselage between the vertical stabilizers. *(Boeing)*

The F/A-18D is operated by several Marine all-weather attack squadrons, including VMFA(AW)-225, shown here. Note the false canopy on the bottom of the forward fuselage. This photograph also shows the size difference between an AGM-88 HARM (on the inboard pylon) and an AIM-7 Sparrow (outboard). *(Robert L. Lawson via the Mick Roth Collection)*

A variety of different paint schemes have been used by the aggressor squadrons. This wraparound scheme was on a Top Gun aircraft (BuNo 162891) at Miramar on 30 March 1996. *(Craig Kaston via the Mick Roth Collection)*

Royal Australian Air Force Hornets from No. 77 Squadron, No. 3 Squadron, and No. 2 OCU fly over Port Stephens, about 100 nautical miles north of Sydney. Note the absence of national markings on the upper surface of the wings. *(RAAF/LAC Jason Freeman via Chris Hake)*

Canadian CF-18s from No. 433 Squadron drop flares durin NATO exercise. Note the false canopy on the bottom of each craft. All CF-18s wear this bit of deceptive camouflage. *(Reyno/Skytech Images)*

This desert scheme was on an F/A-18A (BuNo 162114) assigned to NSAWC in September 1998. Note that only the upper surfaces are camouflaged and the lower surfaces retain their normal gray scheme. *(Mark Munzel)*

Finland calls its aircraft F-18s instead of F/A-18s to distinguish their defensive role. This F-18D (HN-465) shows the overall light gray paint used by all Finnish Hornets. *(Finnish Air Force)*

Malaysia decided to paint its F/A-18Ds in the same dark gunship gray used by the F-15E. The paint is broken only by a small full-color national insignia on the forward fuselage (under the LEX) and a small fin flash on the top of the vertical stabilizer. *(Peter Steinemann/Skyline APA)*

Another operator that prefers the F-18 designation is Switzerland. This F-18D (J-5232) is typical of the Swiss aircraft. Since most Swiss fighters are assigned to a common pool and "borrowed" by individual squadrons as needed, there are no unique squadron markings. All aircraft wear full-color national insignia on the tail and wings. *(Christoph Kugler)*

Spanish EF-18s have routinely been participating in NATO and UN missions around the Mediterranean. The aircraft wear a small full-color roundel on the aft fuselage, but otherwise lack any color. *(Boeing)*

The first EMD F/A-18E (BuNo 165164). Note the clipped-wing AIM-120C on the right middle pylon. *(Boeing photo by Kevin Flynn)*

Canada has painted some unique "specials" over the years, such as this CF-18A (188749) celebrating the fiftieth anniversary of No. 410 Squadron in 1991. *(CAF via the Terry Panopalis Collection)*

The U.S. Naval Test Pilot School operates four Hornets in this very attractive white-with-red-trim scheme. Note the early-design elliptical-cross-section fuel tank on the centerline station of this F/A-18B (BuNo 161249) during May 1989. *(Mick Roth Collection)*

Another aggressor scheme is shown on this VFC-12 F/A-18A (BuNo 162470) at NAS Oceana on 31 August 1995. A careful examination will show some details that attempt to make the aircraft look like a Russian fighter—the black "cut marks" on the tops of the vertical stabilizers, the "air louvers" on top of the LEX, etc. *(Craig Kaston via the Mick Roth Collection)*

The Kuwaiti Air Force decided to camouflage its C/Ds, the only operator to choose a multicolor paint scheme. The color should blend well with the generally clear skies over the Middle Eastern deserts and Persian Gulf that the Hornets operate in. *(Peter Steinemann/Skyline APA)*

Many different techniques are used in wind tunnels to gather data, and instrumentation is one of the largest expenses in preparing models. This F/A-18 model takes a slightly different approach. The pink paint is pressure-sensitive, and actually changes color as the pressure on its surface changes. The model can be photographed with a high-resolution video camera during the test runs, and pressures can be deduced by the color changes. This pioneering work was accomplished at the NASA Ames Research Center. *(NASA)*

The first production-representative F/A-18E (BuNo 165533) shows the different materials used for its skin. The black panels are graphite-epoxy composite, while zinc chromate covers aluminum alloy skin. The radome and air intakes are also of composite construction and are molded with a gray color in the gel coat. *(Boeing)*

The speedbrake has been deleted from between the vertical stabilizers, replaced by two small spoilers located on the upper fuselage on either side of the spine, even with the leading edge of the wing. Speedbrake effect is generated by the ailerons and spoilers going up, the trailing-edge flaps going down, and the rudders deflecting outboard, with the stabilators acting to prevent pitch transients.[160]

The RUG-II APG-73 radar used by late C/Ds is also used in the F/A-18E/F, as are the ASQ-173 LDT/SCAM and AAR-50 TINS pods. An updated ATFLIR version of the NITE Hawk pod will replace the AAS-38B. The radar can process 24 targets and prioritize the top 8 threats. The use of different colors to designate enemy, friendly, and unknown targets aids situational awareness. An interesting piece of information displayed on the radar includes the number of degrees the target has to turn to get away from a fired missile. A recent azimuth track capability added to the radar can differentiate aircraft stacked on top of each other.[161]

The initial ECM suite consists of the ALR-67(V)3 RWR, ALQ-214 integrated defensive electronic countermeasures (IDECM), ALE-50 towed decoy, and four ALE-47 dispensers. Eventually, the ALE-55 decoy will replace the ALE-50. For the time being, the Super Hornet's avionics suite is 90 percent common with the F/A-18C/D, a move intended to minimize development risk and reduce cost. Boeing has, however, reserved 17 cubic feet of available space, excess cooling, and electrical power for future upgrades and new equipment.

The Super Hornet cockpit has been designed for improved situational awareness. The key pad system mounted below the Kaiser HUD in the earlier Hornets has been replaced by an upfront 4×5-inch touch-sensitive liquid crystal display (LCD), which is activated when the pilot touches the display. The touch screen allows the pilot to enter communication channels, navigation waypoints, and other related data. The data entry format also has been structured to lessen entry strokes. The up-front display also can be used to show infrared or radar information. The 5×5-inch central display of the C/D has been replaced by a new 6.25 $\times 6.25$-inch flat panel active matrix LCD, although the two other 5×5-inch color MFDs are retained. This gives the Super Hornet pilot four full displays, plus the heads-up display for combat, performance, navigation, and system information. A separate display is used to

The Super Hornet uses small fairings that space sensor pods away from the fuselage. This is an AAS-46 pod being used on the EMD/LRIP E/Fs until the ATFLIR is developed. The side of the air intake includes an exhaust duct for bypass air. *(Mick Roth)*

The first F/A-18E carries an interesting load—an AIM-7, AGM-65 Maverick, and AGM-154 JSOW under each wing. The arresting hook is attached to a prominent housing under the fuselage. *(Boeing)*

show engine and fuel information. The rear cockpit of the F/A-18F has similar instrumentation, except that it does not have a HUD and the 6.25×6.25-inch screen is located above the landscape-format touch screen, largely mimicking the arrangement in later F/A-18Ds.[162]

A multifunction information distribution system (MIDS), Rockwell digital communication system (DCS), and Hazeltine APX-111 combined interrogator-transponder are installed in the ninth E model (BuNo 165536) and subsequent aircraft. In the near future, a Smiths/Harris tactical aircraft moving map capability (TAMMC) and a Kaiser/Elbit joint helmet-mounted cueing system (JHMCS) will be incorporated. Once the MIDS and TAMMC are installed, the Super Hornet pilot will be able to view full 360° digital map coverage provided by the Northrop Grumman E-2s and other sources.[163]

In March 1997 the DAB approved entering low-rate initial production (LRIP), and perhaps more significantly, delegated the authority to enter full-rate production to the Navy. Usually the full-rate production decision is made by the DAB, further lengthening the approval process. There have been three batches of LRIP aircraft—LRIP-1 (8 Es and 4 Fs in FY97), LRIP-2 (8 Es and 12 Fs in FY98), and LRIP-3 (14 Es and 16 Fs in FY99).[164] The LRIP-1 aircraft were all delivered by 9 November 1999, with the last LRIP-1 aircraft delivered almost 2 months ahead of schedule. LRIP-2 aircraft will be delivered between January and October 2000. It is expected that LRIP-3 will be delivered between November 2000 and July 2001. VFA-122 is the first Fleet squadron to receive the Super Hornet, and began recieving 23 LRIP-1 and LRIP-2 aircraft in November 1999.

These will be followed by full-rate production of 36 (15 Es and 21 Fs) aircraft in FY00 and 42 in FY01 (14 Es and 28 Fs). Production was scheduled to increase to 48 aircraft in FY01, but Congress did not authorize sufficient funds to cover the rate increase. The average unit price (in FY90 dollars) has been $36.4 million.[165] Future production could vary depending on the progress of the Joint Strike Fighter program. Current plans are to procure 548 Super Hornets, but according to the Department of Defense 1997 Quadrennial Defense Review, if JSF deployment is delayed beyond 2008 to 2010, the Navy could purchase as many as 785

Northrop Grumman test pilot Jim Sandberg pauses for a photo opportunity prior to refueling Super Hornet E-1 carrying three 480-gallon drop tanks, two Mk 84 2,000-pound bombs, two AGM-88 HARM missiles, and two AIM-9 Sidewinders. All of the stores pylons are on the fixed part of the wing. *(Boeing photo by Kevin Flynn)*

Super Hornets. The FY00 Defense Appropriations Bill granted the Navy authority for the multi-year procurement of 222 Super Hornets over 5 years (FY00 through FY04). The multi-year contract will save an estimated $700 million.

Boeing and Northrop Grumman have stated that the aircraft was designed completely electronically, using software that automatically tracked all parts changes and ensured the engineers were aware of any incompatibilities introduced by seemingly simple or unrelated changes. Overall the aircraft uses 33 percent fewer parts (11,789, down from 17,210) than the C/D, and fabrication and assembly man-hours are about 12 percent lower. About 27,000 employees of 3,200 companies participate in the Super Hornet program. Northrop Grumman produces and assembles the F/A-18's center and aft fuselage, vertical stabilizers, and all associated subsystems at its Air Combat Systems facilities in El Segundo, California. Forward fuselage manufacturing and final assembly are carried out at the Boeing facility in St. Louis that has assembled all other U.S. Hornets. Aerospace Technologies of Australia (ASTA)[166] delivered the first set of F/A-18E/F trailing-edge flaps in a ceremony at Port Melbourne, Australia, on 26 February 1998. ASTA was the first company outside of North America to be selected as a supplier for the Super Hornet.

Not Stealth, but Reduced Observability

The buzzword among many military planners during the 1990s has been "stealth," or low observability (LO). But there is still a great deal of debate over the merits of stealth technology, especially after the loss of a Lockheed F-117A Nighthawk over Bosnia in 1999.[167] The F-117, and indeed the Air Force's F-22, pay a penalty for their stealth capabilities, shown in their high maintenance requirements and limited capabilities in some areas. It is generally agreed, however, that some aspects of stealth technology are well worth applying to almost any combat aircraft and carry little if any penalty.

There was no way that the F/A-18 configuration could be converted into a stealth aircraft. But there were some technologies that could be applied to it in order to significantly lower its radar signature from certain aspects, particularly nose- and tail-on. It should be noted that similar modifications have been made to almost all U.S. fighters over the past decade.

The F/A-18 modification package was introduced during the production of Lot XII. The gold tint on the canopy of the Night Attack Hornets comes from a thin layer of indium-tin-oxide that serves to reflect radar and laser energy. Although this slightly increases the over-all radar reflection of the canopy (which scatters pretty well because of its curved shape), it keeps radar from reaching the variety of flat surfaces located inside the cockpit that could potentially provide a much sharper and stronger radar reflection. The nose radome is, nat-urally, transparent to most radar frequencies (so the aircraft's radar can operate). But the antenna drive system and the flat bulkhead it is attached to make great radar reflectors. A panel of radar-absorbing material (RAM) was added over the bulkhead and its equipment to absorb radar energy.[168] The AESA antenna being developed for the E/F will eliminate the drive system entirely.

Radar-absorbing paint, containing carbonyl iron particles in a polymer binder, was used to lower the reflectivity of the engine air intake tunnels. This not only absorbs incoming radar signals, but it helps absorb energy reflected back from the face of the engine com-pressor. The F-14 was the first Navy aircraft to be treated in this manner, and the modifica-tions proved quite successful. All of this is not free from penalties, however. Reportedly the RCS reduction effort added 242 pounds to the empty weight of late-model F/A-18C/Ds, further reducing their carrier bring-back capability and range. Also, the original carbonyl iron paint, which was developed for the Air Force, was susceptible to corrosion in sea air. A new paint, developed specifically for the Navy, has largely eliminated this problem.

Boeing's F/A-18 general manager Mike Sears confirms that the F/A-18E/F "is not intended to be an invisible aircraft," but rather a compromise that offers reduced observability with

The designers of the E/F did not get carried away with stealth technology, but they did try to incor-porate items that offered a high payback with relatively little cost. Such details as the diamond-shaped area around the angle of attack sensor in this photo have a rather dramatic effect on reducing the RCS from certain aspects, yet cost little to install or maintain. The blister under the nose contains antennas for one of the ECM systems. *(Mick Roth)*

affordability. Design features that contribute to a lower RCS include downward- and outward-angled engine air inlets that reflect radar returns away from the nose-on aspect, serrated edges on the main landing-gear and engine-access doors, diamond-shaped metal screens covering all apertures, and the minor use of various coatings and other surface treatments.

Of particular interest are the fixed airframe-mounted, radial vanes forward of the engine that prevent radar energy from reaching the rotating fan. This "baffle"[169] is deiced by hot bypass air from the engine. It does cause a small loss in installed thrust, but the overall loss is half of what was predicted, and the baffle actually improves the engine's stall margin under certain conditions.[170]

The number of protruding antennas has also been reduced by using more flush units, and the temperature and pitot/static probes have been combined onto a single stub sensor.[171] The overall RCS of the F/A-18E/F is approximately that of the F-16, and somewhat less than the earlier F/A-18C/D, despite the fact that the new aircraft is 25 percent larger overall. In fact, designers had originally intended to use limited applications of RAM on some exterior surfaces, primarily the leading edges of the wings and empennage. Preliminary testing showed that the aircraft could meet Navy RCS requirements without the use of this material (which is difficult to maintain during operational use), and it was deleted. Overall the E/F carries 154 pounds of RAM, 88 pounds less than the C/D.

Initial Testing

The first EMD F/A-18E (E1; BuNo 165164) made its maiden flight on 29 November 1995 with Fred Madenwald at the controls and was delivered to Pax River on 14 February 1996. This signaled the beginning of a 3-year test program involving over 2,000 flights by seven aircraft. The first test flight was conducted at Pax River on 4 March 1996.[172]

Each of the seven EMD aircraft was assigned a specific task. The first aircraft was used to investigate flying qualities and open the flight envelope. The second F/A-18E (E2; BuNo

The fourth EMD aircraft was assigned to investigate the high-angle-of-attack characteristics of the Super Hornet. Like most modern test programs, safety was a primary concern and the aircraft was fitted with a spin recovery chute between the vertical stabilizers to aid the pilot if necessary. This October 1996 photograph was taken during a deployment to test the parachute. This was the only time the spin chute was deployed in flight—it was never required for spin recovery. *(Boeing)*

This is a slightly unusual attitude to launch an AGM-65 Maverick from—usually the aircraft is diving toward the target. An AGM-88 HARM is under the wing, along with an external fuel tank. The second two-seater (F2) was involved in weapons trials at Pax River, where this photo was taken. *(Boeing)*

165165) made its maiden flight on 26 December 1995 and was delivered to Pax River on 19 February 1996. This aircraft was used primarily for engine and performance testing. The third F/A-18E (E3; BuNo 165167) made its maiden flight on 2 January 1997 and was delivered to Pax River on 1 February to be used for load testing. The fourth single-seater (E4; 165168) made its first flight on 2 July 1996 and was delivered to Pax River on 22 August 1996. This aircraft was allocated to high-angle-of-attack evaluations. The last F/A-1E (E5; BuNo 165170) made its maiden flight on 27 August 1996 and was delivered to Pax River on 25 October 1996. This was the first aircraft equipped with a full mission capability. Weapons systems testing began in October 1996, with stores separation tests beginning on 19 February 1997 with the jettison of an empty fuel tank.[173]

The first F/A-18F (F1; BuNo 165166) made its initial flight on 1 April 1996 and was delivered to Pax River on 21 May 1996. This aircraft was initially assigned to carrier suitability trials. The first three successful catapult launches were made from the land-based catapult at Pax River on 6 August 1996, and initial carrier qualification trials were conducted in January 1997 aboard the USS *John C. Stennis* (CVN-74). Testing lasted 5 days and included 64 arrested landings and launches and 54 touch-and-go landings. The aircraft then performed weapons testing for the remainder of 1997 before being assigned to follow-on sea trials during 1998. The second F model (F2; BuNo 165169) made its first flight on 11 October 1996 and was delivered to Pax River on 23 January 1997. This aircraft was equipped with a full avionics suite and was used for weapons system testing. By 16 November 1996 the F/A-18E/F had flown as high as 49,500 feet, exceeded Mach 1.5, achieved +7/−1.7g, and demonstrated controlled flight at +57/−39° angles of attack.[174]

Initial testing generally went well, but one major problem was discovered in early 1996. An uncommanded rapid change in roll angle, also called "wing drop," occurred in the middle of the maneuvering envelope. During initial flight testing, wing drop was found to occur between 7 and 12° angle of attack at 0.70 to 0.95 Mach. The sudden change in lift was apparently caused by a "turbulent separation in the transonic regime." The phenomenon was "predictable" because it could be expected to occur when flying within this regime; however, it was not "repeatable" because the conditions that caused the onset of wing drop, and the resulting direction and severity of roll, varied. Sometimes the pilot could easily correct the roll with a small bank-angle change; other times, full lateral stick was not enough to "lift" the wing. When this occurred the pilot had to reduce the angle of attack to eliminate the wing drop. Other times the wing drop did not occur at all.[175]

Traditional aerodynamic fixes, such as vortex generators, did not alleviate the problem. An October 1996 change to the flight control software greatly reduced the problem, but did not prove completely successful. Although Boeing and the Navy never doubted they would find an acceptable solution to the problem, the issue exploded in Washington and in the popular press. The Navy was accused of covering up the problem. The GAO called for an immediate suspension of Super Hornet LRIP production until a fix was found, and some members of Congress threatened to cancel the entire program. Neither action occurred. A formal investigation by the Department of Defense Inspector General (IG) concluded on 16 August 1999 that the Navy had not withheld a disclosure of the problem to top Pentagon officials or Congress. The IG's report said, "Overall, Navy officials planned, conducted, and reported operational test results in accordance with the procedures established in DoD regulations and Navy instructions. . . . Further, F/A-18E/F Super Hornet program officials reported the in-flight phenomena known as wing drop in accordance with established risk management and deficiency reporting procedures."[176]

A blue ribbon panel including both NASA Ames and NASA Langley representatives conducted an independent review of the wing drop phenomenon during September 1997. As additional flight testing revealed no easy solutions, a second blue ribbon panel was formed in December 1997 and made numerous recommendations. To better understand the flow mechanisms of wing drop, the panel recommended a flight test program with supporting wind tunnel and computational fluid dynamics (CFD) experiments. Another panel recommendation was that the DoD "initiate a national effort to thoroughly and systematically study the wing drop phenomena" so that future programs could benefit.[177]

A porous fairing over the wing-fold joint was determined to be the most promising fix, and multiple versions were tested to determine the optimum design. The porous wing fairing was believed by many to offer most of the same benefits as a large wing fence, but without its associated problems. The concept was that the higher pressure behind the shock wave toward the trailing edge of the wing would create a downward pressure into the rear of the chamber under the porous fairing. This results in the higher pressure traveling forward in the chamber and exiting upward through the fairing near the leading edge of the wing. This increase in pressure in an otherwise low-pressure area thickens the boundary layer and results in a "virtual fence" without the increase in drag usually associated with a mechanical fence.[178] It should be noted that the final porous fairing was mostly an empirical solution; i.e., it was observed to work in the desired manner, but exactly how it works remains largely unexplained. The description given above is but a single theory among many of how the porous fairing works. NASA Langley has been closely monitoring the progress of the Navy in resolving this issue, and plans to undertake fundamental research to thoroughly understand the problem and its solutions.

A limited number of test flights were conducted on each proposed fairing configuration. With almost all of the proposed fairings, wing drop was found to be greatly reduced, but two forms of buffet were introduced. The first was experienced in 1g transonic flight and was like driving on a gravel road. The other occurred at a variety of altitudes and configurations and was associated with varying angle of attack in a turn. Although neither buffet was considered severe, engineers continued to refine the fairings in an attempt to minimize them. This followed the blue ribbon panel's recommendation to "continue the systematic approach of flight-testing to optimize the design so as to minimize buffet and other impacts."

Further testing was conducted in two phases. In the first phase, 18 sorties were flown emphasizing weapons delivery accuracy and limited low-level flights while equipped with a

The first production single-seater (E6; BuNo 165533) in the markings of the VX-9 operational eval-uation squadron at China Lake. The ailerons droop during low-speed flight, effectively giving the air-craft full-span trailing edge flaps. *(Boeing photo by Dave Martin)*

developmental version of the porous wing fairing. The schedule on the leading- and trailing-edge flaps was also reprogrammed, with the flaps deploying a little earlier and extending a lit-tle farther. These flights suggested that wing drop was controlled to the point that it was relabeled "residual lateral activity," which involved small, uncommanded changes in wing bank angle that were moderate, correctable, and assessed to have little mission impact.

The second phase consisted of a 41-flight interim operational test by VX-9 in mid-1998 that focused on tactics and representative mission profiles. These missions included dis-similar air combat maneuvering, fighter escort, interdiction, and close air support. Two air-craft flew with the production design of the porous wing fairings and the most recent flight control software. Preliminary results over a wide aerodynamic spectrum representative of the tactical environment pointed to a further decrease in residual lateral activity. Some activity remains but has been termed more like a wing tremble. Buffet was still present and remained identical to what was experienced in the developmental test phase flights. Again, it did not interfere with task accomplishment but, during long-duration missions, could become tiring and distracting. Engineers continue to tweak the flight control system in an attempt to further reduce the phenomenon.[179]

Initial flight testing showed that acceleration of the E/F was comparable to that of the F/A-18C in subsonic and negative-*g* regimes. However, the E/F was slower to accelerate to supersonic speeds in 1*g* flight compared to the F/A-18C equipped with the -402 engines. Closely tied to this is aircraft climb performance, which was considered somewhat sub-standard above 30,000 feet. Minor aerodynamic tweaks and flight control law changes have largely eliminated these concerns.

Rear Admiral Johnson said that the testing of the Super Hornet went well, with the only major deficiency being a lateral instability at gross weights above 60,000 pounds at 35,000 feet. When attempting to pull into a slice turn for a weapons delivery, the aircraft tended to roll off in the opposite direction. A thickening ("blunting") of the trailing edge of the flaps and ailerons on production aircraft largely fixed this problem.[180]

The first F/A-18F, piloted by Fred Madenwald with Ricardo Traven in the rear seat, reached the 3,000 flight-hour mark on 1 July 1998 over Pax River. During this time, the Super Hornets cleared the basic flight envelope of the aircraft, conducted more than 500 aerial refuelings, completed initial sea trial carrier qualifications, and evaluated crosswind

The USS *Harry S. Truman* (CVN-75) hosted the second round of sea trials during February and March 1999. Here the first two-seater taxis toward a catapult. Note the faceted shape of the auxiliary air intake door on top of the fuselage. A fuel dump is located immediately above the rudder on each vertical stabilizer, along with various ECM antennas and a position light (on the right side only). *(Boeing photo by Ron Bookout)*

landings. Weapons separation tests included firing 25 missiles and dropping more than 500,000 pounds of ordnance.[181]

A second round of sea trials was conducted during late February and early March 1999 using the two EMD F/A-18Fs aboard the USS *Harry S. Truman* (CVN-75). Captain Robert O. Wirt, Jr., the Navy Flight Test Director for the F/A-18E/F program, said there were no surprises during the test series off the Florida coast. During the tests Lieutenant Commander Lance Floyd made the first night carrier landing with an F/A-18F. On several occasions the test pilots intentionally overshot the centerline of the carrier on turns to final to evaluate the aircraft's maneuverability in making last-minute corrections. The aircraft's response

F1 approaches the *Truman* preparing for a trap. The only external stores being carried are AAS-46 and AAR-50 pods on the fuselage stations, AIM-9s on the wingtip stations, and a centerline fuel tank. Note the asymmetrical deflection of the ailerons and stabilators. The Super Hornet completed this second round of sea trials 3 days ahead of schedule. *(Boeing photo by Ron Bookout)*

The second two-seater also participated in the sea trials aboard the *Truman,* shown here immediately after being catapulted off the angle deck. Note how both rudders toe in toward the center to provide additional nose-up pitch during takeoff. The two boundary-layer dump vents can be seen on top of the LEX, partially covered by the new spoilers that took the place of the dorsal speedbrake. *(Boeing photo by Ron Bookout)*

won praise from both test pilots and landing signal officers who were watching from the stern of the carrier. Partially because of its larger wing, the E/F has approach speeds about 8 knots slower than the C/D.[182]

Other tests included catapult operations off the bow with 15 knots of crosswind, and operations off the angle deck with 10 knots of crosswind. Catapult and landing operations were also conducted with asymmetrical weapons loads. The two aircraft performed automatic carrier landing system approaches from 4 to 8 miles out with a data link to the carrier providing a hands-off ride for the pilot all the way to touchdown. All of these tests were described as "uneventful." Minimum "end speed" tests after catapult launches were conducted in both military power and full afterburner. In afterburner it was expected that the aircraft would leave the deck at 142 knots at its maximum gross weight of 66,000 pounds, and would sink only 10 feet below the bow during its departure. The aircraft passed these tests also.[183]

Testing was not limited to aerodynamic trials, since each of the electronic systems also had to be tested. Because a majority of these were identical to ones found in earlier models of the F/A-18, few problems were encountered. However, an interesting anomaly was discovered with the new ALE-50 towed decoy—the cable burned through under certain flight conditions with the afterburner selected. Flight testing also demonstrated that the heat from the basic engine exhaust (no afterburner) could cause the dielectric material in the cable to soften, thus causing arcing that inhibited electrical power to the decoy.[184]

The preliminary fix for this problem was a "goal post" attachment on the aft bottom of the fuselage designed to hold the cable farther from the jet exhaust. Subsequent evaluations showed that this was an effective fix for the ALE-50. But the ALE-50 is an interim solution pending the development and delivery of the Sanders ALQ-214 IDECM suite, which will use

Like all Navy aircraft, the E/F uses the probe-and-drogue refueling system instead of the Air Force "flying boom" system. Traditionally this has created a problem where Air Force tankers had to be fitted with an adapter to enable them to refuel Navy aircraft. Here is F2 being refueled from an Air Force KC-135R fitted with the MPRS (multipoint refueling system). Boeing rolled out the first of 150 modified KC-135s on 2 March 1997. The modified aircraft have a refueling pod under each wingtip. The pod is an In Flight Refueling Ltd. unit, similar to the French C-135F pods. For safety "wingtip clearance" reasons the boom cannot be used when the tip pods are being used. *(Boeing)*

an ALE-55 fiber-optic tow cable that has different properties from the ALE-50 tow cable. The current IDECM design has the fiber optics wrapped around the outside of the cable's structural core. Flight characteristics and reaction to exhaust heat of the fiber-optic cable are not fully known; however, similar problems with engine exhaust burning the cable are likely. Another unknown factor is the effect of the IDECM fiber-optic cable rubbing or chaffing against the goal post fixture rather than free floating in the air stream. Flight tests of a mass model representing the IDECM decoy have been conducted, and further flight tests of the IDECM design are continuing during late 1999, using a modified F/A-18D.

The overall reliability of the F/A-18E/F proved to be better than expected during the test series, and the aircraft received high praise from the test crews. In addition, controls, displays, and night lighting have received favorable comment from both test and operational pilots.

Most modern combat aircraft are subjected to a "live fire test" program consisting of vulnerability analysis, threat analysis, and ballistic testing of components and major subassemblies. Since it was considered a derivative design, the F-18E/F was granted a waiver to conduct less than full-up system-level testing in May 1992, concentrating on comprehensive ballistic testing of components and major subassemblies instead. This built upon the vulnerability reduction program for earlier F/A-18 aircraft and joint live fire testing of the F/A-18C, as well as actual battle damage incidents.

Live fire testing of the F414 engine indicated that if a blade became separated from a turbine wheel, the blade could penetrate the engine casing and cause damage to other components of the aircraft. The engine casing was redesigned and subsequent tests indicated that separated blades would be contained. Fuel ingestion and ballistic tolerance testing were also conducted on the F414 with encouraging results in comparison to similar tests of the F404.

The F/A-18E/F live fire program includes eight ballistic tests on the third static test airframe (SV52), originally manufactured for drop and barricade testing. Prior to the airframe being assigned to the live fire tests it had been damaged during barricade testing, but had been repaired by Boeing in St. Louis. The airframe (designated DT50 at the time) had been dropped 184 times at higher-than-design loads to evaluate its durability. These included a test at a gross

One of three nonflying F/A-18E airframes constructed by Boeing for various structural and load tests. During this August 1997 test the airframe was shot forward by a catapult just behind the barrier. Note the lack of rudders and stabilators. This airframe was later damaged during a high-speed barrier test. *(Boeing photo by Kevin Flynn)*

weight of 56,000 pounds and a velocity of 28.3 feet per second. A nonballistic fuel cell qualification test was conducted on the airframe during June 1998, which indicated that design goals would be met. Ballistic events against SV52 conducted in late 1999 included testing the stabilator, wing leading edge, engine vulnerability tests, and various fire suppression tests.[185]

Like all military systems, the F/A-18E/F also underwent an extensive year 2000 (Y2K) test to ensure it would not suffer any problems when the new millennium arrived. The testing showed that the Super Hornet program was Y2K-compliant with two exceptions. The first was the tactical mission planning system (TAMPS) that supports various phases of mission planning and employment of the Super Hornet as well as other aircraft. It was not Y2K-compliant when the testing was performed, but a compliant version was released in December 1998. The second area of Y2K-noncompliance was the enhanced comprehensive asset management system (ECAMS) used to track maintenance on the aircraft. This system was replaced in mid-1999 by the automated maintenance environment (AME), which is Y2K-compliant.[186]

Between May and November 1999, VX-9 conducted the formal operational evaluation (OPEVAL) of the E/F, consisting of over 700 sorties. In late August the squadron took the E/F to its first Red Flag exercise at Nellis AFB, flying interdiction sorties, fighter escort, and defense suppression missions. Upon the completion of OPEVAL the squadron will prepare a detailed report on the operational effectiveness of the Super Hornet for senior Navy officials.[187]

SHARP

In mid-1999 the Navy began development of a shared reconnaissance pod (SHARP) for the two-seat F/A-18F. SHARP is expected to restore the capability lost when TARPS-equipped F-14s are replaced by the F/A-18F, and the Navy wants to have the SHARP capability in place for the first Super Hornet deployment in 2003. "That's a short time to develop a pod and get it fielded," said Susan Wright, the Navy's SHARP program manager. The goal is to have 2 to 4 pods ready for the deployment, and this means the Navy will have to rely on nondevelopmental hardware and commercial off-the-shelf items.[188]

SHARP is being designed as a medium- and high-altitude, fully digital reconnaissance system that uses electro-optical and infrared sensors. The system is designed to allow F/A-18Fs to collect imagery from as far away as 45 nautical miles in slant range. Competitors are being given an option of bidding a sensor that can operate at medium and high altitudes (2,500 to 40,000 feet), or two sensors—one for medium altitude (2,500 to 20,000 feet) and one for high altitude (20,000 to 45,000 feet). If two separate sensors are proposed, the Navy wants both to be carried simultaneously. The sensor selection will be made in February 2001.

Raytheon is building the pod that will house the sensor, reconnaissance management unit, data storage system and other equipment during the development phase. The pod, a modified C/D-model 330-gallon external fuel tank, will be carried on the centerline station. The maximum weight for the package is 370 pounds.

The prototype pod will begin flight testing in June 2001 while the Naval Research Laboratory is testing candidate SHARP sensors on a P-3. The formal development program will overlap with the prototype effort, and the engineering and manufacturing development phase will run from FY00 to FY02. A full-rate production decision will not be made until 2004.

The Navy has requirements for 50 SHARP systems for F/A-18Fs, but plans to buy only 22 because of funding constraints, Wright said. That quantity includes six sensors and five pods bought during the EMD phase, and four pods and sensors bought as low-rate initial production items. Some SHARP efforts are not currently funded. The most notable is an upgrade to allow the system to do all-weather reconnaissance using synthetic aperture radar, a capability already fielded with the Marines' ATARS reconnaissance system on the F/A-18D. SHARP would use SAR data from the APG-73 radar that would be relayed into the pod.

The first F/A-18F during a stopover at the Dryden Flight Research Center (note the Space Shuttle training aid in the background). Like earlier Hornets, the F/A-18E/F includes a built-in access ladder that retracts from the undersurface of the LEX. *(Tony Landis)*

The two-seat Super Hornet will likely find other uses in the Fleet as the years go by. Currently Boeing is investigating using the basic F-model airframe as a replacement for the Grumman EA-6B Prowler electronic warfare aircraft. This involves some challenges, mainly workload considerations based on a two-man crew, but offers the benefits of higher performance and commonality with the fighter-attack configurations. *(Tony Landis)*

F/A-18C2W

During 1993, the Navy awarded McDonnell Douglas a small study contract to evaluate the potential of using a modified F/A-18F to replace the Grumman EA-6B Prowler. This was after budgetary constraints prompted the Navy to cancel its planned AdvCap[189] upgrade to the EA-6B. Subsequently, the Navy launched its less costly EA-6B ICAP-3 program as an interim measure. Since the Navy-funded study ended, Boeing has continued its study of a "command and control warfare" (C2W) version of the F/A-18F because of its potential operational advantages for the Navy and an expanded market for the Super Hornet. Northrop Grumman, which developed the EA-6B and the Air Force's EF-111A counterpart, joined the effort in 1996 as electronic warfare (EW) system integrator. The ongoing effort has been funded entirely by the companies.[190] Navy interest in the concept was piqued in 1999 during the air campaign over Yugoslavia, when more than 50 of the 90-some EA-6Bs were tasked to support those operations.[191]

The basic C2W configuration adds wingtip pods containing ICAP-3 wideband receivers, with ALQ-99 jammer pods carried on the four inboard wing pylons. This would allow the two outer wing stations to carry AGM-88 HARM missiles and a centerline fuel tank. AIM-120s could be carried on the fuselage stations for self-defense. A hump behind the cockpit would contain satellite communications equipment. Because the ALQ-99 pods have not been tested at supersonic speed, and were not designed to withstand high-g maneuvers or minimize their RCS, they may need to be modified for use on a C2W version of the F/A-18F. This version is also referred to as the *airborne electronic attack* (AEA) aircraft, "F/A-18G," or "Growler," although none are official designations.[192]

Boeing has built a cockpit simulator in St. Louis to refine the cockpit configurations and perform crew workload studies. This is especially important since one of the primary drawbacks to the Air Force EF-111A was its two-member crew versus the four-member crew on the

As part of the predevelopment activities for the possible F/A-18G "Growler" electronic warfare aircraft, Boeing conducted fit-tests at Pax River using ALQ-99 jammer pods borrowed from the EA-6B community. Note the pods on the wingtips in place of the normal Sidewinder rails. Here the first two-seat F/A-18F shows an AGM-88 HARM on the outboard pylon, an ALQ-99 jammer pod on the middle pylon, and an AGM-154 JSOW on the inboard pylon. An AIM-120 AMRAAM is on the fuselage corner station and another ALQ-99 pod is on the fuselage centerline. *(Boeing photo by Kevin Flynn)*

The F/A-18G Growler is not an approved program, and faces some tough challenges. One is whether a two-seat aircraft can adequately perform the mission; the EA-6B takes four crewmembers, and having only two crewmembers was one of the reasons given for retiring the EF-111A Ravens. Here the first F/A-18F is shown with a different load: ALQ-99 pods on the fuselage centerline and the inner and middle wing pylons, and a HARM on the outer wing pylon. *(Boeing photo by Kevin Flynn)*

EA-6B. Boeing points out that processing technology has improved dramatically over the past few years, allowing the computer to perform many of the functions once done manually.[193]

In July 1999 Boeing briefed Rear Admiral John Nathman, Director of Air Warfare, and Rear Admiral Jeffery Cook, program executive officer for tactical aircraft, on the status of its studies. Boeing indicated that the first C2W versions could be available by 2006 if work began quickly.

In November 1999 General James Jones, the Corps Commandant, suggested the Marines were looking more seriously at the proposed "Growler" as a temporary replacement for its EA-6B Prowlers that must be retired by 2015. The Marines want a jamming aircraft based on the Joint Strike Fighter (JSF), but one will not be ready by the 2010 Pentagon deadline for selecting an EA-6B follow-on. Jones said that flying the Growler until a JSF derivative is available is an option under consideration: ". . . [A]dopting the F/A-18G temporarily is a possibility. We're working our way through the issues." A version of the F/A-18F would not only benefit from logistics commonality with the rest of the air wing, but would have comparable flight performance, which would make it easier to integrate into a strike group.

Boeing continues to investigate a two-seat F/A-18F equipped with a version of the ICAP-3 electronics being developed for the EA-6B. Boeing and the Navy have been debating the development costs of the Growler. Boeing officials said company and Navy estimates are within 20 to 25 percent of each other, a figure that is considered small at this point in a project. During mid-1999 Boeing conducted fit-checks at Pax River using the first F/A-18F and a set of ALQ-99 pods. The F/A-18F already has sufficient excess electrical power to run any new electronics that are not self-powered with ram-air turbines. Initial electromagnetic interference tests indicate that the Super Hornet's fly-by-wire flight control system is sufficiently hardened to resist any interference from the new jammers.[194]

A go-ahead decision is not expected before 2002.

The Science: Construction and Systems

The F/A-18 is a carrier-capable, twin-engine all-weather fighter and attack aircraft. The aircraft has been manufactured in two major variants (A/B/C/D and E/F), with single-seat (A/C and E) and two-seat (B/D and F) versions of each. The United States Navy and Marine Corps have been the major customer, with export versions sold to Australia, Canada, Finland, Kuwait, Malaysia, Spain, and Switzerland.

Although the E/F is essentially a different aircraft from earlier Hornets, the descriptions that follow are generally applicable to both designs. In cases where there are notable exceptions between versions (or export customers, for that matter), they have been explicitly stated.

Fuselage

The fuselage is a conventional semimonocoque structure constructed primarily of aluminum alloys, with a titanium firewall located between the engine bays. Some steel and titanium alloys are also used in high-strength portions such as landing gear bulkheads. The primary skin material is aluminum, with graphite-epoxy composite used for most access doors, landing gear doors, and engine bay doors. Electroluminescent formation lights are located on the sides of the forward fuselage. The radome is a filament-wound fiberglass-epoxy shell that is electrically transparent to radar signals. There are stores stations on each fuselage corner just aft of the engine air intake, and a single centerline stores station rated to carry up to 2,600 pounds. The centerline station is plumbed to accept fuel tanks.[195]

The forward fuselage contains the single M61A1 20-mm cannon, which fires through an opening on top of the fuselage just behind the radome. A variety of antennas are mounted on the forward fuselage. On some aircraft there are five short blade antennas (called "bird slicers") just ahead of the windscreen for the APX-111 CIT system. On each side of the forward fuselage there are antenna blisters for the ALQ-165 forward upper high-band transmitter (above the formation light strip), ALR-67 radar warning receiver

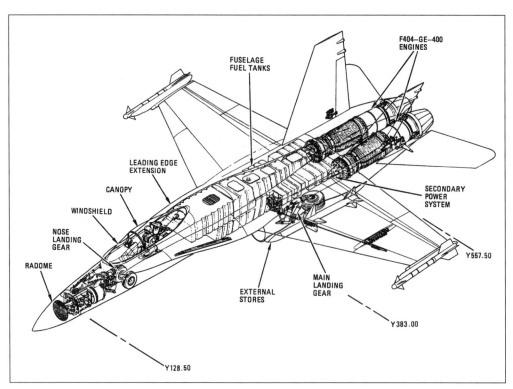

This overall arrangement diagram shows the major components of the F/A-18A, but all models are generally similar. Most of the center fuselage is taken up with fuel tanks, while the aft fuselage contains the engines and other power systems. The forward fuselage contains the avionics, pilot, and 20-mm cannon. *(U.S. Navy)*

(below the light strip), and the ALQ-165 forward receivers (just below the leading edge of the LEX). On the bottom of the forward fuselage is an antenna blister for the ALQ-126B/165 forward lower high-band transmitter (on the gun door), five stub antennas for the ALR-67 low-band array (just aft of the blister), ALQ-165 forward lower low-band transmitter (on the nose landing gear door), and an ultrahigh-frequency/very high frequency (UHF/VHF) blade antenna. The dorsal spine has a blister on each side just behind the cockpit for the ALQ-165 forward upper low-band transmit antennas, and two blade antennas, the forward one being used for the ARN-118 tactical air navigation (TACAN) equipment and the other for the ARC-182 UHF/VHF radio. There is a blister on the underside (slightly inboard) of each air intake for the ALQ-126B antennas (low-band on the left; high-band on the right), and a flush fairing on the aft end of each corner stores station for more ALQ-126B antennas (again, low-band on the left; high-band on the right). U.S. Navy C/Ds carry either the ALQ-165 or no jamming equipment, depending upon where they are based. The blisters remain on the aircraft even when the jamming systems have been removed. The A/B models do not carry ALQ-165 and are not equipped with the antennas unique to that system. This may change if either Australia or Canada elects to upgrade its aircraft with ALQ-165.

On Hornets, a twin-hinged hydraulically activated speedbrake is mounted on the upper rear fuselage, between the vertical stabilizers, providing minimal pitch change when the speedbrake is extended. The Super Hornet deletes the speedbrake, but adds a pair of small spoilers located on the upper fuselage on either side of the spine, even with the leading edge of the wing. Speedbrake effect is generated by the ailerons and spoilers going up, the

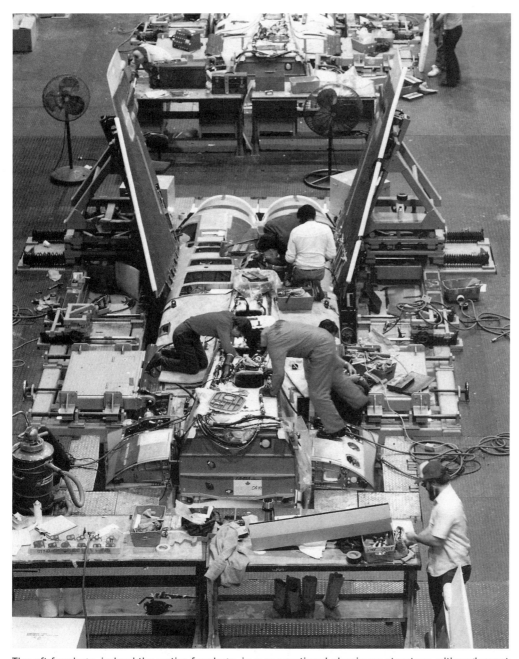

The aft fuselage, indeed the entire fuselage, is a conventional aluminum structure, although most of it is covered by graphite-epoxy composite doors. The radical angle of the original P530's vertical stabilizer was reduced to only 18° for the YF-17 and F/A-18. *(Tony Landis Collection)*

trailing-edge flaps going down, and the rudders deflecting outboard, with the stabilators acting to prevent pitch transition.

Cockpit

Each crewmember is provided with a Martin-Baker[196] Navy Aircrew Common Ejection Seat (NACES) in a pressurized, heated, and air-conditioned cockpit. The original (BuNos 161353 through 164068) Martin Baker SJU-6/A (front seat) and SJU-5/A (rear) were

The Hornet's dorsal speedbrake is mounted well aft, between the vertical stabilizers. The Super Hornet deletes this speedbrake, choosing instead to use small spoilers installed on top of the LEX, plus other control surfaces to perform the speedbrake function. *(Mike Reyno/Skytech Images)*

All Hornets have a boarding ladder that retracts into the underside of the LEX. Note the standoff support that extends to the fuselage. The boarding ladder can be operated only from outside the aircraft. This Swiss F-18 also shows the large spotlight fitted to some export Hornets. *(Neil Dunridge)*

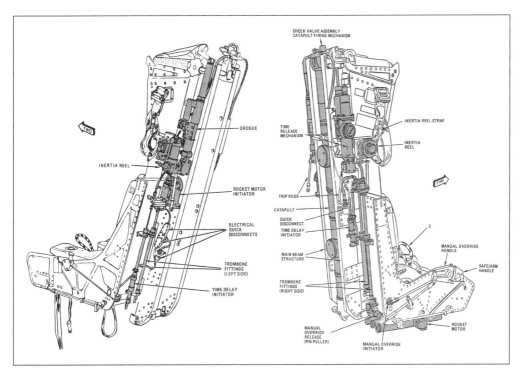

The Martin Baker NACES ejection seat is common to most Navy aircraft, greatly simplifying logistics and training. *(U.S. Navy)*

superseded (since BuNo 164196) by the SJU-17-1/A (front) and -2/A (rear).[197] The seat is capable of zero-zero operation (i.e., it can provide meaningful escape while sitting stationary on the ground, or after a "cold cat" launch failure aboard carrier). The cockpit maintains a pressure altitude of 8,000 feet up to an actual altitude of 23,000 feet. Above this the pressure altitude steadily increases to approximately 14,500 feet at 35,000 feet, and 20,000 feet at 50,000 feet actual altitude. The cockpit is accessed via a retractable boarding ladder stowed under the left LEX. Ladder retraction can only be accomplished manually from outside the aircraft.[198]

The upward opening clamshell canopy is made of stretched acrylic attached to an aluminum frame, and there is a fixed, single-piece, forward windscreen. The two-seat canopy is 54 inches longer and 125 pounds heavier than the single-seat canopy. When closed, the canopy is secured by three latches and a forward latch pin on each side. Night Attack variants of the C/D, and all E/Fs, have a gold-tinted (actually a thin layer of indium-tin-oxide) canopy to deflect radar and laser energy.[199]

The F/A-18B and the first 31 F/A-18Ds were intended primarily as proficiency trainers and retain a full set of flight controls (stick, throttle, rudder pedals) in the aft cockpit. The Night Attack D and F models carry a weapons systems operator in the rear cockpit, which is not normally equipped with flight controls, although it is possible to fit flight controls to the aft cockpit of D and F models for use as a trainer.

Primary cockpit instrumentation includes a Smiths Industries multipurpose color map display in both cockpits, two Kaiser multifunction displays (monochrome in early aircraft; color in Night Attack Hornets and the E/F), a central multipurpose GEC-Marconi-AlliedSignal display, Kaiser AN/AVQ-28 heads-up display, and a GEC-Marconi FID-2035

The wing-fold hinge is a very robust structure to absorb the loads encountered in flight. The leading-edge flaps are split at the hinge line, while the hinge also separates the ailerons from the trailing edge flaps. Unlike many U.S. aircraft, the Hornet uses large external fairings to cover the actuator mechanisms for the ailerons and flaps. This F/A-18C is carrying an AGM-154 JSOW on the outboard pylon. Note the folding wings on top of the JSOW. *(U.S. Navy)*

This is the LERX from the YF-17 demonstrator, showing the large boundary layer air discharge (BLAD) slots between the LERX and fuselage. These slots had the beneficial effect of generating a strong vortex extending down each side of the fuselage, increasing directional stability at high angles of attack. The first F-18s were also delivered with these slots. *(Dennis R. Jenkins)*

Unfortunately, the BLAD slots also generated a great deal of parasitic aerodynamic drag, which adversely affected range and acceleration. Consequently, 80 percent of the length of the slots was filled in beginning with the eighth FSD F-18A, leaving only one small slot on each side whose function was to eject the boundary layer air bled from the engine intake. *(Dennis R. Jenkins)*

horizontal situation display. Night Attack Hornets include GEC-Marconi MXV-810 Cat's Eyes night vision goggles.[200]

Wings

The cantilever midwing of the Hornet has a trapezoidal planform (swept on the forward edges, but straight on the trailing edges) with 3° anhedral and 20° sweepback at quarter-chord. The multispar construction uses primarily aluminum alloys with graphite-epoxy interspar skin panels and trailing-edge flaps. The full-span leading-edge flaps have a maximum extension angle of 30°, while the single-slotted trailing-edge flaps deploy to a maximum 45°. Ailerons are located on the outer wing panel and can be drooped to 45°, effectively creating full-span trailing-edge flaps. The leading- and trailing-edge flaps are under computer control and deflect to create the optimal combination of lift and drag under all flight conditions, including air combat maneuvering and cruise conditions. The ailerons and flaps are deflected differentially for additional roll authority as necessary. All flight control surfaces are hydraulically actuated and commanded electrically by the flight control system. Each wing contains an integral 96-gallon fuel cell. Large LEXs are of aluminum construction and permit flight at angles of attack exceeding 60°. The outer wing panels fold upward 100° for storage using a Garrett AiResearch mechanical drive. Electroluminescent formation lights are located on small vertical plates extending above and below the wingtips. Each wing is equipped with a wingtip rail station for AIM-9 Sidewinder missiles and two wing stations rated at 2,600 pounds for other stores. The inner station is plumbed for an external fuel tank. In addition to the stores stations included on earlier Hornets, the E/F has one additional station under each wing rated to carry 1,150 pounds. The inboard and middle stations on the Super Hornet are plumbed for fuel tanks.

Empennage

The swept vertical and horizontal surfaces are constructed primarily from graphite-epoxy composite skins over a light alloy honeycomb core. The vertical stabilizers are canted outward at 20° to ensure they are in clean air at high angles of attack, and are mounted farther forward than usual, with their leading edges overlapping the trailing edge of the wing. The rudders are deflected inward during takeoff to provide improved nose wheel liftoff characteristics and shorten the takeoff roll. The rudders are also toed in at low angles of attack, and as the angle of attack is increased (during rotation, for instance), the toe-in decreases and even begins to flare out at higher angles of attack. In this manner, rudder toe/flare contributes to the aircraft's longitudinal pitch characteristics.

The stabilators have 2° of anhedral and are activated collectively for pitch control and differentially for roll control. The rudders and stabilators are all hydraulically actuated under command from the flight control system. Electroluminescent formation lights are located on the outer surface of the vertical stabilizers. On the Super Hornet, a damaged or failed stabilator will automatically return to a fixed position and the flight control system reverts to different control laws to provide near-normal flight characteristics using the remaining flight control surfaces.

The three trailing-edge fairings on the F/A-18C/D's left vertical stabilizer are (from top) an ALQ-126B/165 high-band transmit antenna, an ALR-67 receive antenna, and an ALQ-165 low-band receive antenna. On the right stabilizer they are a position light, an ALR-67 receive antenna, and ALQ-165 low- and high-band receive antennas. The ALQ-165 antennas are not functional on aircraft not equipped with the system. The A/B models do not have the ALQ-165 antennas.

This VFA-151 F/A-18C shows most of the upper-side details of the Hornets. The speedbrake can be seen between the vertical stabilizers, as can the small slot on top of the LEX that remains to dump bleed air from the engine intake. The markings here are fairly typical of Navy and Marine aircraft, with only a small low-visibility national insignia on the left wing. *(Mark Munzel)*

Landing Gear

The retractable tricycle-type landing gear has a twin-wheel nose unit and single-wheel main units. The nose unit retracts forward under the cockpit, and both nose gear doors remain open while the gear is extended. The main units retract rearward, rotating 90° to stow horizontally[201] inside the lower surface of the engine air ducts. The main gear is equipped with an antiskid system. The nose gear is equipped with a combination shimmy

The trailing edge of the A/B model vertical stabilizers had fewer antennas than later C/D models. On the left side there were (from top) an ALQ-126 high-band transmit antenna, an ALR-67 receive antenna, and a fuel dump vent. The right side had a position light, an ALR-67 receive antenna, and another fuel dump vent. These are the early small fairings—on later aircraft the fairings extended a great deal farther forward and were larger in diameter. There were also position lights on each side of the vertical, immediately above a fuel system vent. A formation light strip was positioned vertically. This VFA-132 aircraft was photographed on 18 August 1984 before the aluminum dou-blers were added at the vertical stabilizer attach points. *(Mick Roth Collection)*

The C/D added another antenna on the trailing edge of the vertical, and changed the shape of the ones that had been there. On the left side (from top) are an ALQ-126B/165 high-band transmit antenna, an ALR-67 receive antenna, and an ALQ-165 low-band receive antenna. On the right stabilizer there are a position light, an ALR-67 receive antenna, and ALQ-165 low- and high-band receive antennas. The ALQ-165 antennas are not functional on aircraft not equipped with the system. Note the small doublers at the vertical stabilizer attach points. *(Mick Roth)*

damper and dual-mode nose wheel steering system. Steering is controlled by a switch on the control stick and hydromechanically operated through inputs from the rudder pedals and flight control computers. During low-speed taxi the nose wheel may pivot ±75°, while during high-speed taxi the limit is ±16° to prevent overcorrection. The nose wheel steering system is disengaged when power is removed from the aircraft, allowing ground crews to

The twin nose wheel on the Hornet is arranged in a much more robust design than the unit used on the YF-17 demonstrators. The launch bar can be seen in its raised position on the front of the strut of this Swiss aircraft. One of the UHF/VHF blade antennas can also been seen here. *(Neil Dunridge)*

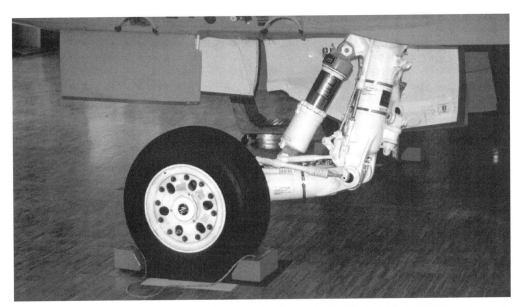

The main gear of the Hornet is also much more robust than that of the YF-17, a necessary measure to ensure the 24 feet per second sink rate requirement for carrier landings was met. When retracted, the wheel will rotate 90° to lay flush against the underside of the air intake duct. Note the AIM-7 Sparrow mounting location just above the landing gear of this Swiss F-18C. *(Neil Dunridge)*

tow the aircraft. The E/F model incorporates yaw-rate feedback that helps the aircraft track straight on the runway.

The nose unit uses $22 \times 6.6\text{-}10$, 20-ply tires pressurized to 350 psi for carrier operations and 150 psi for shore operations. The main units use $30 \times 11.5\text{-}14.5$, 24-ply tires pressurized to 350 psi for carrier operations and 200 psi for shore operations. A carrier-capable arresting hook is installed under the aft rear fuselage of all Hornets.

The launch bar is in its down position, attached to a catapult on an aircraft carrier. Only the United States routinely operates Hornets off carriers, and Australia deleted the launch bar to save some weight and complexity. Note the landing light on the upper part of the nose gear strut. This F/A-18 is assigned to VFA-113 aboard the USS *Constellation. (U.S. Navy/DVIC DN-SC-86-01305)*

The E/F model uses essentially the same design, just strengthened as appropriate to accommodate the heavier gross weight of the aircraft.

Engines

F/A-18A/B/C/D

The A/B and early C/D models are equipped with two General Electric F404-GE-400 turbofans rated at 10,600 lbf dry and 15,800 lbf with afterburning. Later C/D models (beginning at Block 36 in January 1991) use two F404-GE-402 enhanced performance engines (EPE) rated at 12,200 lbf dry and 17,600 lbf with afterburning.[202]

Compared to other recent turbofans, the F404 experienced few developmental problems and demonstrates great resistance to compressor stalls, even at high angles of attack. The engine is very responsive, with spool-up from idle to full afterburner in just 4 seconds. The engine control system regulates speeds, temperature levels, and fuel flow for afterburning and nonafterburning operation. The lubrication and ignition system are self-contained on the engine. The engine is continuously monitored for critical malfunctions and parts life usage by an in-flight engine condition monitoring system (IECMS).

Overall maintenance statistics for the F404 are excellent. The engine has required less than two shop visits per 1,000 flight hours. The in-flight shutdown record is even more impressive, averaging 6,500 hours between events.[203] This means that a typical pilot could spend an entire flight career without experiencing an in-flight shutdown.[204]

The engines in the F404 series are dual-rotor, augmented, low-bypass turbofan engines. They are modular engines that incorporate a three-stage fan and a seven-stage high-pressure compressor, each driven by a single-stage turbine. The combustor is a through-flow annular type, and the afterburner is fully modulating from minimum to maximum thrust. The F404 is 159 inches long and 34.5 inches in diameter. The -400 engine weighs 2,161 pounds, while the -402 weighs 2,237 pounds. The -402 has a slightly lower bypass ratio (0.27 versus 0.34), but a high compression ratio (27:1 instead of 25:1) and a higher fan pressure ratio (4.3 versus 3.9). This results in the -402 having a thrust-to-weight ratio of nearly 8:1 while the -400 is rated at 7:1 (approximately).[205]

The three-stage fan (low-pressure compressor) is driven by a single-stage turbine. Approximately one-third of the fan discharge air is bypassed to the afterburner for combustion and cooling. The seven-stage high-pressure compressor is also driven by a single-stage turbine. The first- and second-stage compressor stators are variable, and fifth-stage compressor air is used by the engine anti-ice system. A set of variable inlet guide vanes is mounted in front of both the fan and compressor to direct the inlet air at the best angle for engine operation. Atomized fuel and compressor discharge air are mixed and ignited in the combustion chamber, then passed through the compressor and fan turbines and out the engine exhaust.[206]

Afterburner operation adds more atomized fuel, mixed with the combustion discharge gases and bypass fan discharge air, to produce additional thrust. The electrical control assembly, variable exhaust nozzles, main fuel control, and afterburner fuel control provide coordinated operation of the engine throughout its flight envelope. The engine accessory gearbox, driven by the compressor rotor, powers the lubrication and scavenge oil pumps, variable-exhaust-nozzle power unit, alternator, main fuel pump, and afterburner fuel pump. An aircraft-mounted auxiliary power unit is used to start the engines.

The F404-GE-402 was developed in response to a Swiss requirement for additional power, and Kuwait subsequently specified the engine because the desert's hot environment typically reduces a jet engine's performance significantly. After reviewing the performance increase afforded by the new powerplant, the United States decided to introduce it on late-model F/A-18C/Ds. The -402 is essentially the same engine as the -400, but uses single-crystal alloys in both turbine stages and higher-temperature alloys in the compressor.[207]

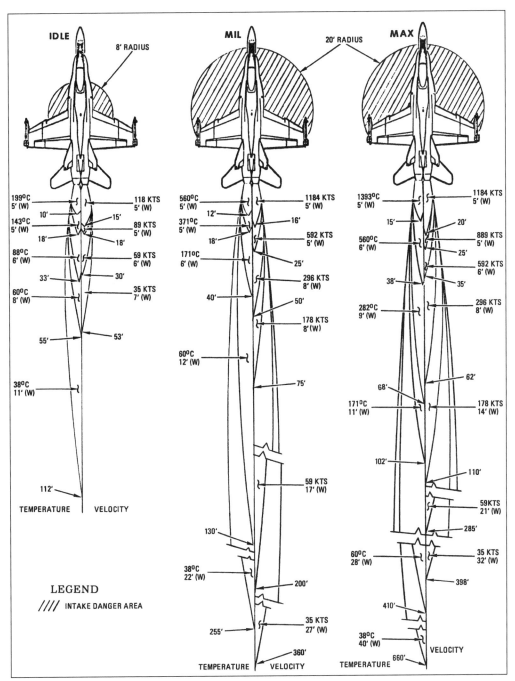

This diagram shows the danger areas associated with the engine intakes and exhaust. At MIL power the exhaust is moving at 1,184 knots and its temperature is over 1,000°F. At MAX (afterburner) power, the exit velocity has not increased, but the temperature has more than doubled. Things are much calmer at IDLE, although caution is still needed near the aircraft. *(U.S. Navy)*

F/A-18E/F

The Super Hornet is equipped with two General Electric F414-GE-400 turbofan engines rated at 14,770 lbf dry and 21,890 lbf with afterburning. The new engine is the result of a systematic program of F404 developments that began in 1983. Under this program, GE built and tested a variety of improved fans, cores, and combustors suitable for F404 derivatives, and this technology was used in the F412-GE-400 medium-bypass nonafterburning turbofan developed for the A-12. After the A-12 was canceled, this research was applied directly to an engine for a larger F/A-18 derivative.[208]

The F414 uses the basic F412 core, which had undergone 4 years of full-scale development and testing, combined with the advanced low-pressure system from the YF120 engine developed for the Advanced Tactical Fighter competition (YF-22 and YF-23). Noteworthy features include an advanced afterburner using ceramic-matrix composite materials and a robust one-piece "blisk" fan with a 17 percent higher airflow than the F404. The full-authority digital engine control (FADEC) system developed for the F412 is also used. The engine does not share any major parts with the F404.

The F414-GE-400 consists of six engine modules and an accessories package. The engine incorporates a three-stage fan driven by a single-stage low-pressure turbine and a seven-stage axial-flow high-pressure compressor driven by a single-stage high-pressure turbine. The combustor is a through-flow annular type, and the hinged flap cam-linked exhaust nozzle is hydraulically actuated. The engine-mounted accessory gearbox provides the necessary extracted power needed to drive the accessories. The engine control system regulates speeds, temperature levels, and fuel flow for afterburning and nonafterburning operation. The lubrication and ignition system are self-contained on the engine. The engine is 155.5 inches long, has an inlet diameter of 30.6 inches, and weighs 2,445 pounds. The E/F uses a "throttle-by-wire" control system, and unlike earlier Hornets there is no mechanical linkage between the throttle and the engine.[209]

The first two production F414s were delivered in ceremonies on 12 August 1998 at GE's Lynn, Massachusetts, facility. These were the first of 28 LRIP F414-GE-400s, and 13 addi-

The air intakes on the early-model Hornets were simple D-shape inlets with nonmoving ramps or variable-geometry features. This was because there was no requirement to exceed Mach 2 with the aircraft. Two small white plates just behind the inlets are covering the ALE-39/47 countermeasure dispenser locations, and the nearest technician is reaching into the main gear well. (Tony Landis Collection)

tional engines were delivered by the end of the year. By the end of 1998 the 21 EMD engines had accumulated 6,300 flight hours since the first F/A-18E began flying in November 1995, as well as an additional 6,100 hours in ground tests. Twelve development test engines had accumulated about 12,600 hours of factory tests.[210]

A follow-on LRIP contract brought the total engines on order to 142. Final LRIP deliveries are to be made by the end of May 2001, with F414 full-rate production expected to follow the next month at a rate of two engines per month. This should increase to six, then ten engines per month by 2002, according to George Bollin, GE's F404/F414 project manager.[211]

Air Intakes

The D-shaped intakes on the original F/A-18s and the trapezoidal intakes of the F/A-18E/F are set well back underneath the LEXs, which guide air into the engine intakes and somewhat protect them from high-angle-of-attack airflow disruptions. Since the F/A-18 is not required to exceed Mach 2, the aircraft does not need sophisticated variable-ramp air intakes, using instead a simple, fixed splitter plate mounted next to the fuselage. Operating manuals contain a warning that airframe buffet and vibration due to engine inlet duct rumble might occur above Mach 1.75. The only moving parts are two ducts cut into the top of the LEX that eject bleed air upward into the LEX-generated airflow. The ducts automatically open at Mach 1.33 during acceleration and close at Mach 1.23 while decelerating. The rear part of the compression ramp is porous to prevent boundary layer air from entering the inlet. Part of the boundary layer air is bled through a fixed-area outlet into the fuselage boundary layer diverter channel and under the fuselage, while the remainder exits via spill ducts atop the LEX.[212]

Fuel System

F/A-18A/C

Total internal fuel is 1,597 gallons (10,381 pounds), housed mostly in four main fuel tanks (containing 426, 249, 200, and 530 gallons) in the swollen dorsal spine. These tanks are installed in a row beginning just behind the cockpit and ending just forward of the engines. No fuel is stored between the engines. Each wing also contains a 96-gallon integral fuel cell. Fuel is always used from any external tanks first, then the wing tanks, and finally the fuselage tanks.

All tanks and fuel lines are self-sealing, with reticular foam in the wing tanks (to suppress possible fire and/or explosion). The soft inner surface closes after projectile penetration, but is less effective against punctures any larger than relatively small-caliber rounds or shrapnel (like most SAM-induced damage). The fuel feed lines in the main gear wells are also wrapped with a self-sealing shell. Tank baffles provide a limited fuel supply during negative-*g* flight, but there are no provisions for sustained negative- or zero-*g* flight. Later F/A-18C/Ds incorporated a computer "square balancing" system that monitors and evenly distributes fuel throughout the aircraft, freeing the pilot from the tedium of manually managing fuel balance.

The retractable probe for the in-flight refueling system is positioned in a flush housing on the right side of the fuselage just forward of the windshield. Provisions are made to carry three 330-gallon external fuel tanks, one on each inboard wing station and one on the centerline fuselage station, all on SUU-62 adapters. Empty tanks may be carried on the outer wing pylons if necessary, but no plumbing exists at these stations. A 480-gallon tank external tank is also available that can be carried on all three stations, but is not cleared for carrier use on the F/A-18A/B/C/D. Canada and Kuwait both use the larger tank as standard

The Hornet, like all Navy aircraft, uses the probe-and-drogue method of aerial refueling. This F/A-18 is carrying an AGM-88 HARM missile on the inboard pylon. *(Mick Roth Collection)*

equipment. Single-point refueling is via a connector on the left side of the forward fuselage. Fuel dumps and vents are located at the top of the vertical stabilizers.

During initial introduction of the F/A-18A, McDonnell Douglas developed a conformal external dorsal tank that would have added only 300 pounds of empty weight while providing an additional 3,000 pounds of fuel. No customer expressed an interest in the idea at

Like most modern aircraft, the Hornet is equipped with single-point refueling, allowing all six fuel tanks (plus any external tanks) to be refueled from a single location on the aircraft. This Swiss F-18 shows the location on the left side of the aircraft, where it will not interfere with pilots climbing out of or into the aircraft. The location is also far enough rearward to allow ground crewmembers to reload the 20-mm cannon, but far enough forward to avoid the inrushing air around the air inlets if the engines happen to be running. *(Neil Dunridge)*

This CF-18A is moving forward to engage the drogue. Note the large spotlight on the forward fuse-lage just under the tip of the LEX. This is used to identify unknown aircraft during night intercepts, and was installed on all Canadian, Swiss, and Kuwaiti aircraft. The Canadians also use a "false canopy" on the bottom of their aircraft as part of their camouflage. *(Mike Reyno/Skytech Images)*

the time, and it was dropped without being prototyped. In 1996–1997 McDonnell Douglas tested the viability of replacing the external tanks with larger 660-gallon tanks. Like the 480-gallon tanks, these are not cleared for carrier operations and the United States has not purchased any. They are available to export customers but none are known to have purchased any.

The Royal Australian Air Force uses KB707 tankers equipped with pods on each wingtip to refuel their AF-18s. This is the view from the back seat of a No. 77 Squadron AF-18B during refueling operations from a KB707 from No. 33 Squadron. *(RAAF/LAC Peter Gammie via Chris Hake)*

F/A-18B/D

The two-seat models carry only 316 gallons in fuselage tank 1 (instead of 426 gallons). Otherwise the system is identical to the single-seater.

F/A-18E/F

These aircraft are generally similar to earlier Hornets, except total internal fuel capacity is increased to 2,130 gallons (14,500 pounds). A new type of fuel tank, originally developed for the F-15E, is being used on the E/F models. Made of a thinner, more flexible polyurethane, the new bladders are lighter, more flexible, and easier to install than the Nitrol rubber units used earlier. The tanks in older F/A-18 models are being replaced with tanks of the new material on an as-needed basis, and new-production C/Ds are using the new material.

Each Super Hornet can carry up to five 480-gallon external tanks manufactured by Advanced Technical Products Inc.'s Lincoln Composites division. Because of the flexible quantity of tanks carried by the Super Hornet, it is anticipated that between 700 and 1,000 fuel tanks will eventually be ordered. With an air-refueling system (ARS) pod, the Super Hornet can be used for aerial refueling of other aircraft. The 480-gallon fuel tank uses an inner fuel permeation barrier and outer structural shell of filament-wound composites. The inner layer consists of fiberglass impregnated with an epoxy resin specifically formulated for fuel compatibility. The outer layer consists of a carbon-fiberglass hybrid impregnated with another epoxy resin and is the primary structural component of the tank.

Other Systems

The Hornet features a single airframe-mounted accessory drive (AMAD) just forward of and below the engines. Hydraulic pumps, generators, fuel pumps, and air starter turbines are all mounted on the AMAD, which is connected to each engine via a drive shaft. The AMAD greatly simplifies maintenance since it is not necessary to disconnect all of these systems in order to change an engine. All that is required is disconnecting the appropriate drive shaft and removing the engine while leaving all of the other systems undisturbed. Another advan-

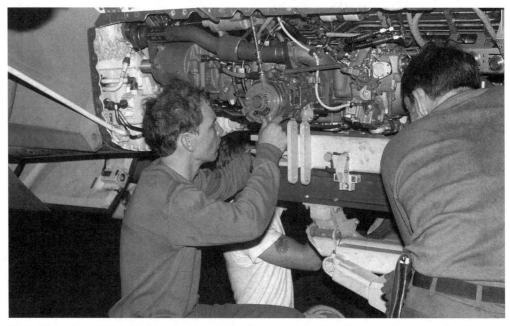

Technicians from VFA-113 disconnect accessories in preparation for removing an engine from an F/A-18A aboard USS *Constellation*. (U.S. Navy/DVIC DN-SC-86-01292)

tage is that the drive shaft can be disengaged on the ground by pulling a lever that allows the maintenance crew to power up the hydraulic system, electrical system, or fuel system without running an engine. The auxiliary power unit (APU), or a ground starter, can spin the air turbine starter, which will run these systems without turning the engine.

Two completely separate 3,000-psi hydraulic systems are each rated at a maximum flow rate of 56 gallons per minute. The E/F uses 3,000/5,000-psi hydraulic systems with the pressure changes commanded as a function of airspeed. The hydraulic systems provide power for all flight controls, the speedbrake, the refueling probe, landing gear and brakes, and the M61 cannon. A sophisticated fire detection and extinguishing system is installed in the engine compartments. A single AlliedSignal GTC-200 APU is provided in the A/B/C/D models for engine starting and ground pneumatic, electric, and hydraulic power.[213] An upgraded APU is installed in the Super Hornet.

As a point of interest, it should be mentioned that the F/A-18C has 307 access doors, 90 percent of which can be reached from the ground without work stands.[214]

Electronic Systems

In the original Hornets, a General Electric quad-redundant digital fly-by-wire (FBW) control augmentation system (CAS) provides the commands to the flight control actuators, with direct electrical backup to all surfaces, and a direct mechanical backup to the stabilators. The FBW system receives inputs from the pilot via the control stick and rudder pedals, computes which control surfaces to deflect to accomplish the intended movement, then electrically commands the hydraulic actuators at the appropriate surfaces. Under normal circumstances the FBW system will not allow the pilot to overstress the airframe, although the pilot may opt to override if necessary.[215] Since the 14 November 1980 crash of an F/A-18 (BuNo 161215) into Chesapeake Bay (the pilot lost control at 20,000 feet and had to eject), all Hornets have been outfitted with a spin recovery mode (SRM). The SRM, whether engaged automatically or manually with a switch, does not apply antispin or dive recovery controls—that is left up to the pilot. What SRM does do is give the pilot full lateral control authority to arrest the yaw rate, thereby breaking the spin. Once the yaw rate is reduced, it is up to the pilot to fly the aircraft out of the subsequent nose-low attitude.

The Super Hornet uses a very similar flight control system except the mechanical (Mech) backup linkage to the stabilators has been eliminated.[216] Even on the original F/A-18, many McDonnell Douglas engineers suggested that the Navy not specify a mechanical backup system. The Mech system added mechanical complexity and weight, most of which is aft of the center of gravity. However, the engineers' strongest argument was that the incorporation of a Mech backup would require additional failure-sensing hardware and software logic for both stabilators, all of which would have to be four-channel redundant. Their feeling was that the failure-monitoring hardware itself was susceptible to failure and might cause reversions to the Mech mode even though the flight control system was working as designed. Since one of the original design goals for the Hornet was increased reliability, McDonnell Douglas was attempting to minimize the hardware complexity of the flight control system. The A/B/C/D also has poor raw longitudinal stability characteristics, making it very difficult to fly in the Mech mode. However, since this was the first digital FBW aircraft to enter production, the Navy was opposed to the idea of deleting all mechanical backup systems, and the Mech was retained for the A/B/C/D.

When development of the E/F began it was obvious that the aircraft would be even less longitudinally stable than the previous versions. Whereas it is very difficult to fly the A/B/C/D in the Mech reversion mode, it would be nearly impossible to do so in the E/F. Additionally, the Navy had gained a great deal of operational experience with the F/A-18's flight control system and was much more comfortable deleting the Mech linkage altogether.

The flight control system incorporates departure resistance features in the normal (CAS) flight control logic. While maneuvering at elevated angles of attack, the flight control system (while in CAS, not SRM) will wash out the pilot's lateral control commands so as to limit the amount of yaw rate the pilot can generate, thereby helping to prevent a nose slice departure. This feature of lateral control washout is incorporated in all versions of the F/A-18. McDonnell Douglas had proposed an automatic spin recovery logic for the E/F which would automatically apply antispin controls. The Navy was not prepared to accept this, and elected to remain with the logic developed for previous versions of the Hornet. The E/F advances the antideparture logic even further and provides for the flight control system to actually apply small increments of antiyaw lateral controls to counter a beta (sideslip) buildup, with no input from the pilot.

Most F/A-18s are equipped with a Collins ARN-118(V) tactical air navigation (TACAN) system. The system provides an output to the INS/GPS system to provide backup position information. A single blade antenna is mounted on top of the fuselage just behind the cockpit.[217]

The Collins ARC-182 VHF/UHF radio is used on most Hornets. The system can access 11,960 channels in the AM, FM, and UHF bands, with 28 channels preselected at any given time. A KY-58 secure speech/TSEC system is used to provide secure communications. Two blade antennas are used, one on top of the midfuselage, and one under the forward fuselage.

An AlliedSignal AN/APX-100 IFF set was originally fitted to all Hornets. The Hazeltine AN/APX-111 combined interrogator transponder (CIT) was specified by Kuwait for its aircraft and has been selected by the United States to retrofit all U.S. Hornets. It is likely that most other operators will follow suit. The APX-111 is readily identifiable by the row of five short antennas on top of the forward fuselage ahead of the windscreen. The APX-111 features electronic beam steering that allows it to determine the range, bearing, and relative elevation of any aircraft when interrogated.[218]

U.S. aircraft are fitted with an automatic carrier landing system (ACLS) for all-weather carrier operations. With this system, the carrier's landing system radar (AN/SPN-42) tracks the aircraft and compares the aircraft position with the optimum approach profile. The aircraft position is then corrected to the desired glide path by commands from the Navy Tactical Data System (NTDS) over a UHF data link to the aircraft's AN/ASW-27B/C data link receiver. The data link receiver then directs pitch and roll commands to the flight control computers. In addition to automatically controlling the aircraft, the system can transmit and display discrete messages on the pilot's instrument panel. The ACLS enables completely hands-off landings onto aircraft carriers from a final path location roughly 8 miles away. The carrier's SPN-42 also transmits azimuth and elevation glide slope signals to the ASW-27, which are sent to the Telephonics Corporation AN/ARN-128(V) receiving decoder set and displayed on a standard crossed-bar indicator in the cockpit. The ARN-128(V) replaced the earlier ARA-63A beginning in 1990 on most aircraft.[219]

The ACLS was deleted from all export Hornets and replaced in most by a commercial instrument landing system (ILS). For the Finnish Hornets this was the Telephonics Corporation tactical ILS, which interfaces with portable units set up alongside various roads during dispersal maneuvers. Other operators have specified a variety of different ILS units.

The original Litton AN/ASN-130A inertial navigation system (INS) was replaced by an AN/ASN-139 integrated laser-ring gyro/GPS unit on all U.S. and most export aircraft beginning in 1991.[220] The Litton ASN-139, also known as the Carrier Aircraft Inertial Navigation System (CAINS II), uses the same form factor as the ASN-130 but substitutes laser-ring gyros for the electromechanical units used previously. The accuracy specification is better than 1 nautical mile per hour drift and 3 feet per second velocity, with a 4-minute reaction time. Aircraft built since September 1995 also have a p-code GPS receiver integrated with the ASN-139. The GPS receiver has been retrofitted to most earlier aircraft.

The joint helmet-mounted cueing system (JHMCS), developed in a Boeing, Lockheed, Air Force, and Navy program, is designed to be installed on the F-15, F-16, F/A-18, and F-22. The Joint Strike Fighter program will also use a variant of the system. The prime subcontractor is Vision Systems International, which is a joint venture between Elbit Systems Ltd., which has designed a similar system for the Israeli Air Force, and Kaiser Electronics. The first flight on an F/A-18 was made on 2 November 1998 at China Lake and lasted 1.7 hours. It examined fit and comfort issues, display stability during high-*g* maneuvers, and accuracy of the magnetic head tracker. Flight testing is scheduled to continue for 14 months. Initial production contracts for F/A-18 JHMCS units include 44 systems in FY00 and 44 in FY01.[221]

The JHMCS combines a magnetic head tracker with a display projected onto the pilot's visor, giving the pilot a targeting device that can be used to aim sensors and weapons wherever the pilot is looking. With JHMCS, the pilot can aim the radar, air-to-air missiles, infrared sensors, and air-to-ground weapons merely by pointing his/her head at the target and pressing a switch on the flight controls. Additionally, the pilot can view selected data (airspeed, altitude, target range, etc.) while "heads-up," eliminating the need to look into the cockpit. The high off-boresight system (HOBS) is the combination of JHMCS and AIM-9X, and can be employed without maneuvering the aircraft, minimizing the time spent in the threat environment. The JHMCS can also cue the pilot to look at something the sensors have detected.

The JHMCS is fundamentally different from the sights used on the MiG-29 and Su-27 because target and seeker information can be displayed on the helmet-mounted display. Radar or infrared targets can be marked on the display, with cueing symbols that direct the pilot's eyes to the target. Tests have shown that a JHMCS-cued pilot will consistently see another aircraft at a greater range than an unaided pilot.[222]

Radar

The liquid-cooled Raytheon (Hughes) AN/APG-65 multimode pulse-Doppler radar has 8,000 fewer parts than the F-4 Phantom's radar, yet offers 20 percent greater range and much-improved reliability. The radar occupies 4 cubic feet of space and weighs 340 pounds, not including the antenna. The set is provided with built-in test equipment, which assists in identifying and isolating failures. According to Raytheon, this "provides total end-to-end radar preflight checkout and continuous monitoring." For maintenance, the radar can be pulled forward and out of the nose on fixed rails once the radome has been opened 180° to the right via its hinge point.[223]

The APG-65 has been superseded by the more capable Raytheon AN/APG-73. Like the APG-70 and APG-71 units used in later F-15s and F-14s, the APG-73 capitalizes on the great advances made in processing power and large-scale integrated circuits in recent years. With no corresponding increase in weight, this radar is 3 times faster, has more memory than its predecessor, is more reliable, and provides for easier maintenance. The resulting unit uses the same antenna and transmitter as the APG-65, mated with entirely new electronics. The receiver/exciter unit provides much faster analog-to-digital conversions, and the new radar data processor replaces two separate units (the signal processor and the data processor) in the old radar. The signal processing speed has been increased almost tenfold. A new power supply that uses solid-state techniques provides a cleaner and more reliable power source. The physical form factor is identical for the two units.

The APG-73 is standard in most late production F/A-18C/Ds, including those for export customers, and is also standard in the F/A-18E/F. In addition, some earlier F/A-18Cs have

The APG-65 radar is a remarkably compact unit. This particular radar is installed on the T-39D test aircraft, not in a Hornet. A careful examination will reveal labels on most of the black boxes warning that they are heavy and require two people to lift them. *(Hughes Aircraft via the Terry Panopalis Collection)*

The APG-73 radar used in the C/D and E/F looks similar to the earlier APG-65, but is a remarkably more capable unit. Here the unit has been slid forward on its built-in rails. Note the configuration of the cannon ports above the radar, and how they blend with the radome when it is closed. *(Raytheon Systems)*

been retrofitted with the unit; the APG-65s removed from these aircraft have been installed in Marine Corps AV-8B+ aircraft, giving them a radar capability for the first time. Canada and Australia both have programs under way to replace the APG-65 in their A/B models with RUG-II APG-73 units.[224]

Several operating modes are available—in general the descriptions are applicable to both radars. The velocity search mode is used for maximum-distance encounters, sacrificing detail for range. This provides only velocity and azimuth information for targets at ranges up to 80 nautical miles. The software controlling the radar is programmed to give a higher priority to those contacts with positive closure rate. In the range-while-search (RWS) mode, the radar provides information on all contacts ahead of the aircraft at ranges between 40 and 80 nautical miles. The APG-73 also has a raid assessment mode, with a range of 35 nautical miles. It uses Doppler beam sharpening to examine a specific return more closely to distinguish between a single target or a group of aircraft in close formation.

While the radar is operating in the RWS mode, a single target track (STT) mode can be selected by the pilot to display steering commands and weapons launch data on the HUD. The system also provides a SHOOT cue when it obtains a firing solution.

The track-while-scan (TWS) mode is employed at ranges up to 46 nautical miles. In TWS mode, the system can track up to ten targets simultaneously and will display the eight deemed most threatening. The computer displays the target's aspect, altitude, and velocity. If the pilot initiates the AIM-7 launch sequence while in TWS, the radar will switch to STT before missile launch. However, obtaining an STT from TWS takes a second or two and is not always successful. For this reason, most (probably all) Hornet pilots manually switch to STT well before missile launch. The transfer from RWS to STT is much faster and more reliable that the transfer from TWS to STT.

When both the target aircraft and the Hornet are maneuvering, there are several automatic acquisition modes available. Once a target has been selected for attack, the system uses the boresight mode if the Hornet is in a traditional tail-chase encounter with the enemy. In this mode, a very narrow 3.3° beam scans a small area of sky directly ahead of the aircraft, providing very accurate range and velocity information. The gun director mode is employed at ranges under 5 nautical miles. The radar provides target position, range, and velocity data to drive the gun aiming point on the HUD. The pilot then positions the pipper on the selected target and squeezes the trigger. The system also has a heads-up display acquisition mode where the radar scans a box corresponding to the field of view of the HUD itself. This typically extends 10° left and right of centerline, 14° above, and 6° below. In the vertical acquisition mode, the radar scans an arc 5.3° wide and extends 60° above boresight axis to 14° below. In order to achieve automatic lock-on, the pilot rolls the aircraft into the same plane of motion as the target, ideally positioning the enemy just above the canopy bow and aligned vertically with the HUD. Since the radar is scanning a large section of airspace, the range and velocity information is slightly degraded. This is less the case for the APG-73, since it has faster processors. Each of these modes results in an STT.

The real beam ground mapping mode is used for identifying substantial geographical features ahead at long ranges. A small-scale radar map displays a representation of this terrain. The computer automatically adjusts the display so that it appears as a vertical "God's-eye" view rather than the oblique view that the radar actually sees. The APG-73 also has more detailed mapping modes that employ Doppler beam sharpening for higher navigation and target location resolutions.

Once a ground target is identified, the air-to-surface ranging mode provides information on the distance to target. It is best employed in a steep dive. Fixed and moving ground target track modes use two-channel monopulse angle tracking to provide precise information on ground targets. There is also sea-surface mode, where the computer filters out clutter to simplify the detection, identification, tracking, and attacking of surface vessels or low-flying aircraft.

Sensors

Various external pod-mounted sensors may be carried for ground attack missions, especially at night. The fuselage corner intake stations, usually occupied by either AIM-7s or AIM-120s when the aircraft operates in the fighter role, are used to carry these pods. The LAU-116 adapters on these stations, which eject missiles away from the aircraft, are usually removed[225] when the sensor pods are carried. Three different pods may be carried, depending upon mission requirements.

AN/AAS-38

Development of the Lockheed Martin (Loral) NITE Hawk FLIR began in 1978 specifically to support the F/A-18. The pod is carried on the left fuselage corner station. It enhances night attack capability (often in conjunction with AVS-9 NVGs) by providing real-time thermal imagery, which is displayed on one of the cockpit displays. The FLIR can be fully integrated with the other avionics of the F/A-18 and data from it can be used in the calculation of weapons release solutions. The FLIR operates in either a narrow (3°) or wide (12°) field of view. The AAS-38A version of this pod is also capable of target designation for laser-guided munitions delivery and are usually referred to as laser target designator/ranger (LTD/R) or targeting FLIR (T-FLIR). The AAS-38B model adds a laser spot tracker that allows the system to search for, acquire, and track targets that are being laser-designated by other aircraft or ground troops. Systems have been delivered to all operators except Finland, which did not procure any since its Hornets are used exclusively for defensive operations. The pod is 6 feet long and 13 inches in diameter and weighs either 348 pounds (AAS-38) or 370 pounds (A/B). The advent of the AAS-38B has largely eliminated the need to carry the ASQ-173 LDT/SCAM pod.

AN/AAS-46

This is a modified AAS-38B pod used on EMD and LRIP E/F models to test T-FLIR functionality during initial testing and operations. This pod will be replaced by the advanced targeting forward-looking infrared (ATFLIR) pod on operational aircraft.

ATFLIR

The Raytheon ATFLIR being developed for the F/A-18E/F made its first flight in November 1999. The ATFLIR pod/adapter features both navigation and infrared targeting systems, incorporating third-generation and midwave infrared (MWIR) staring focal plane technology. ATFLIR has substantially greater recognition range, long-range laser designation, and geopoint accuracy to support precision targeting of GPS-guided weapons. Development was begun in 1996, and the initial operational capability is planned for mid-2002. This pod will functionally replace the AAS-38 on the E/F models, and eventually on the C/D fleet as well. As of the end of 1999 the new pod did not yet have an official designation.[226]

AN/AAR-50

The Raytheon (Hughes) thermal imaging navigation set (TINS) is usually associated with Night Attack Hornets, and is mounted on the right fuselage corner station. It is a fixed FLIR with no designation or tracking capability and is therefore used only for night navigation purposes. The sensor has a 19.5° field of view and displays raster video on the HUD. The system is a derivative of the AAQ-16 installed on various Army helicopters and is also known as the navigation FLIR (NAVFLIR). The pod is 6.5 feet long and 9.9 inches in diameter and weighs 213 pounds with its adapter.[227]

The AAR-50 navigation FLIR is normally mounted on the right fuselage station of Night Attack Hornets. This photo gives a good indication of the size of the pod, and the fairing that is installed on the fuselage station to mount the pod. *(Raytheon Systems)*

AN/ASO-173

The Lockheed Martin (Martin Marietta) laser detector tracker/strike camera (LDT/SCAM) was developed from the Air Force's Pave Penny pod. It is used for accurate bombing in bad weather during daylight, and has no night capability. This pod does not have the capability to designate targets, and is used to identify targets illuminated by either ground troops or other aircraft. Target position data are fed into the F/A-18 mission computer and used to provide weapon-aiming and ordnance release information. The laser optics are stabilized by using attitude data from the aircraft's INS. A KB-35A 35-mm strike-recording still picture camera in the rear of the pod aids in bomb damage assessment. It has a 35° field of view and pans 180°. The pod is 7.5 feet long and 7.9 inches in diameter and weighs 161 pounds.[228] This pod has been largely replaced by the AAS-38A/B.

Additional pods may be carried on the centerline stations for specific missions, such as the AWW-9B and AAW-13 advanced data link pods for the Walleye I/II extended-range data link (ERDL) and SLAM/SLAM extended-range (ER) missiles. The Common Data Link pod is used by F/A-18D(RC) aircraft to downlink reconnaissance data.

Electronic Countermeasures

AN/ALQ-126B

This Sanders (Lockheed Martin) deception jammer is a multimode, power-managed, reprogrammable defensive ECM system. The system responds to the threats by attempting to confuse radars operating at 2 to 16 GHz, which translates to E-band through part of J-band. Available jamming techniques include main-lobe blanking, inverse conical scanning, range-gate pulloff, and swept square waves. Power output is over 1 kilowatt per band at a 4 percent duty cycle. The ALQ-126B uses antennas under the leading edge of the air intake (low-band on the left, high-band on the right), on the bottom of the fuselage on the cannon door, in the trail-

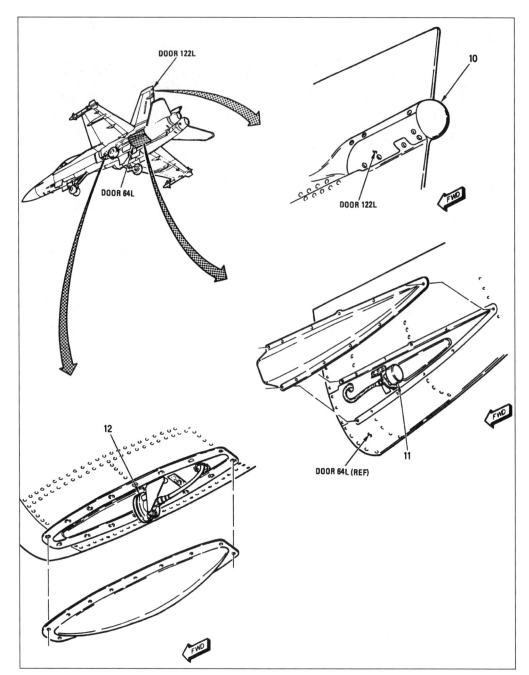

The blisters that house the various ECM antennas are just plastic fairings that cover the actual antennas. This drawing shows the ALQ-126 antennas on the F/A-18A. The blister in the lower left is mounted under the air intake, while the other fuselage antenna is mounted in a fairing on the aft part of the fuselage corner stores station. *(U.S. Navy)*

ing edge of each fuselage weapons station fairing (low-band on the left, high-band on the right), and on the trailing edge of the left vertical stabilizer.[229] The ALQ-126B is installed on U.S., Australian, Canadian, and Spanish F/A-18A/Bs.[230] The ALQ-126B was installed on early U.S. F/A-18C/Ds, but has been replaced by the ALQ-165 when the aircraft are in high-threat areas.

AN/ALQ-162(V)

Development of the Northrop Grumman ALQ-162 continuous wave (CW) jammer was initiated in 1979 to produce a small reprogrammable radar jamming system. The ALQ-162 is designed to intercept and process CW signals, and upgrades under evaluation include responses to pulse-Doppler radars. Jamming signals are produced in various formats to counter threat signals. The system is operated via a control panel located in the cockpit and can be used for stand-alone operation or in conjunction with more sophisticated RWRs such as the ALR-67. It can also be integrated with the ALQ-126 jammer and the ALE-47 countermeasures dispenser. The system is installed on Canadian and Spanish Hornets to supplement the ALQ-126B and provide an all-threats capability.[231]

AN/ALQ-165

Development of the Consolidated Electronic Countermeasures Company (a joint venture between Northrop Grumman[232] and ITT Industries) advanced self-protection jammer (ASPJ) was initiated under a $376 million contract in August 1987 to replace the ALQ-126 as the Navy's standard jamming system. The highly automated, software-controlled system incorporated many state-of-the-art technologies including microwave monolithic circuits and gate arrays and has the ability to automatically select the best jamming techniques to use against any given threat on the basis of the system's own analysis of data received from the RWR. The system can be reprogrammed on the flight line to accommodate changing threat scenarios. The system completed initial verification testing in July 1991, but was canceled when it failed to pass its final operational test in 1992. In retrospect, most believe that the tests were too demanding.

Low-rate initial production of the ALQ-165 resulted in 96 sets being delivered before production was halted, and these were placed in storage at a Navy depot in Indiana. When the Navy became concerned about the sophistication of the air defenses over Bosnia, 24 of the

An interesting paint job on the nose of this F/A-18C still shows most of the ECM antennas mounted on the forward fuselage. On each side there are antenna blisters for the ALQ-165 forward upper high-band transmitter (above the formation light strip), ALR-67 radar warning receiver (below the light strip), and the ALQ-165 forward receivers (just below the leading edge of the LEX). On the bottom of the forward fuselage is an antenna blister for the ALQ-126B/165 forward lower high-band transmitter (on the gun door), five stub antennas for the ALR-67 low-band array (just aft of the blister), ALQ-165 forward lower low-band transmitter (on the nose landing gear door), and a UHF/VHF blade antenna. *(Tom Chee via the Mick Roth Collection)*

sets were removed from storage and installed on Marine Corps F/A-18Ds during 1995. These proved to be much more capable than other available equipment, so another 30 sets were removed from storage and issued to two F/A-18C squadrons when operating in high-threat areas. These systems are not permanently installed on aircraft—they are traded between squadrons as needed to cover deployments to high-threat areas.

ALQ-165 antennas are located on each side of the forward fuselage above the formation light strip (forward upper high-band transmit), on each side of the fuselage under the LEX leading edge (forward high- and low-band receive), on each side of the dorsal spine behind the cockpit (forward upper low-band transmit), on the nose wheel (forward lower high-band transmit) and gun bay doors (forward lower low-band transmit), and on the trailing edge of both vertical stabilizers (low- and high-band receive on the right; low- and high-band transmit on the left).

New-production versions of the ALQ-165 are included on aircraft delivered to Finland, Kuwait, Malaysia, and Switzerland. In August 1996 the Navy issued contracts to cover the procurement of 36 systems to equip additional F/A-18 aircraft, and further procurements were done in FY98 and FY99.[233]

AN/ALQ-214

The Sanders (Lockheed Martin) integrated defensive electronic countermeasures (IDECM) program is producing 548 units at a total cost of $1,930 million, including research and development costs. The average unit cost is expected to be $1.3 million. Sanders is the prime contractor, with ITT Industries being a major subcontractor. In November 1997, the IDECM program was rebaselined to fund an 87 percent development cost overrun and extend the development schedule by at least 6 months. Initial flight tests of the ALQ-214 are scheduled to begin in late 1999.[234]

The IDECM contributes to the *Joint Vision 2010* concept of full-dimensional protection by improving individual aircraft probability of survival. The IDECM suite is intended to provide self-protection for tactical aircraft against radio-frequency (RF) and infrared (IR) surface-to-air and air-to-air threats. The major hardware component to be developed by the IDECM program is the ITT radio-frequency countermeasures (RFCM) system (in other words, the jammer). But perhaps even more important, the IDECM is producing an integrated system that will link other ECM systems together with the RFCM. For the F/A-18E/F these systems include the ALR-67(V)3 RWR, AAR-57 common missile warning system, AN/ALE-47 dispenser, ALE-55 fiber-optic towed decoy (FOTD), and the RFCM.[235]

The ITT RFCM provides a "techniques generator" on board the aircraft, which determines an optimum signal to counter an attacking missile. This is sent via an optical fiber in the tow cable to the ALE-55, where it is amplified and broadcast. The ITT techniques generator consists of three line replaceable units (LRU)—a receiver, a signal processor, and a modulator. The three units collectively weigh less than 150 pounds.[236]

A slightly different version of the ALQ-214 is a possible candidate to upgrade some F/A-18C/Ds. Because F/A-18C/Ds do not carry towed decoys, if they are retrofitted with the ALQ-214, ITT would provide high-band and low-band jamming transmitters that would radiate via antennas on the aircraft. The transmitters would be functionally identical to the units for the ALQ-165.

AN/ALR-67

The Litton/ATI ALR-67 digitally controlled, reprogrammable radar warning receiver uses complex sorting algorithms to compare possible threats with stored characteristics of known radars. The system provides the operator with a digital frequency readout, plus displayed symbology to provide warnings of both ground-based and airborne emitters. The ALR-67 is capable of detecting and processing multiple threat signals and provides the rel-

The ALR-67 antenna on the bottom of the nose presents an odd appearance. The blister just ahead of it houses an ALQ-126/165 lower high-band transmit antenna. *(Mick Roth)*

ative bearing to each signal source with respect to the aircraft. The advanced special receiver (ASR) version was an upgrade of the original ALR-67 that covers an expanded range of frequencies and has improved sensitivity. The ASR variant also has a shorter reaction time, and is capable of processing more signals in a denser threat environment. The ALR-67 uses four receive antennas, one on either side of the nose under the formation light strip and one on the trailing edge of each vertical stabilizer, and an array of five small stub antennas on the bottom of the fuselage immediately behind the ALQ-126 antenna on the gun door. Originally the ALR-67 was used on the F/A-18A/B, being replaced by the ALR-67(V)2 in 1990. The ALR-67(V)3 is manufactured by Raytheon (GM-Hughes), GEC-Marconi, and Litton and uses the same antennas but features a faster processor. This version is better able to handle dense threat environments, particularly ones with frequency-hopping radars, and is better able to detect and track pulse-Doppler and monopulse radars.[237]

The (V)3 is undergoing operational evaluation, which is expected to conclude late during 1999. A low-rate initial production (LRIP) contract for 20 of the systems was awarded on 30 June 1998, with initial system deliveries scheduled for mid-2000. On 18 August 1999 Raytheon was awarded a $62 million contract for 34 ALR-67(V)3 systems to equip the E/F. The (V)3 will be integrated into the Sanders IDECM system on the E/F.[238]

Current plans also call for procuring approximately 150 of the ALR-67(V)3 systems for temporary use on F/A-18C/Ds when they are assigned to high-threat areas. Potential upgrades to enhance the performance of the (V)2 systems have also been developed that would increase sensitivity and reduce ambiguities in identifying the threat radar. A detailed development program will be launched if funds become available in FY00.[239]

AN/ALE-39

This Lockheed Martin (Goodyear) expendable countermeasures dispenser can accommodate RR-129/RR-144 chaff, Mk 46 Mod 1C or MJU-8/B flares, and AM-6988A expendable jammers. Each cartridge is 1.4 inches in diameter and 5.8 inches long. The dispenser is also

Early Hornets, such as this No. 77 Squadron AF-18A, were equipped with two ALE-39 expendable countermeasures dispensers on the bottom of the fuselage just behind the leading edge of the air intake. Here a flare has just been released. Later aircraft have either two or four of the "smarter" ALE-47 dispensers. *(RAAF/LAC Jason Freeman via Chris Hake)*

compatible with the GEN-X expendable jammer developed by Texas Instruments. All A/B models and early U.S. C/Ds used two 30-tube ALE-39 dispensers, one underneath each air intake. On U.S. aircraft these were replaced by ALE-47s. Upgrade programs in Australia and Canada will eventually replace the ALE-39s on those aircraft with ALE-47s.[240]

AN/ALE-47

The Tracor Aerospace expendable countermeasures dispenser is replacing most ALE-39 units on early F/A-18s. The ALE-47 uses the same form factor as the ALE-39 and also uses the same expendable cartridges. The ALE-47 uses threat data received from other sensors to assess the threat situation and determine the appropriate countermeasures response. The system has been included on Hornets delivered to Finland, Kuwait, Malaysia, and Switzerland, and on U.S. aircraft delivered after FY96. Each Hornet carries a single ALE-47 under each intake; the Super Hornet carries two under each intake. In order to increase the total countermeasures load, Boeing has tested a pair of conformal fairings that resemble ventral fins, each containing three 30-tube ALE-47 dispensers. As far as is known, these have not been installed on any aircraft to date.[241]

AN/ALE-50

The Raytheon advanced airborne expendable decoy (AAED) consists of a launch controller, launcher, and decoys. The system was originally developed for use on the A-6E before that aircraft was retired. The ALE-50 can be manually operated as a stand-alone device, or it can be integrated and controlled by the ALE-47. A triple dispenser is located under the fuselage just aft of the landing gear wells on the F/A-18E/F. The ALE-50 decoy is a repeater jammer that is towed behind the aircraft; it is not retrievable after it has been deployed and must be jettisoned before the aircraft lands.[242]

AN/ALE-55

The Sanders fiber-optic towed decoy (FOTD) is the primary jammer associated with the ALQ-214 IDECM system. ITT Industries provides the RFCM "techniques generator" on board the aircraft, which generates an optimum signal to counter an attacking missile. This signal is sent via an optical fiber in the tow cable to the ALE-55, where it is amplified and broadcast. To facilitate the transition from ALE-50s to ALE-55 towed decoys, the Sanders FOTD can be launched from the same aircraft dispensers. Raytheon is developing the improved multiplatform launch controller (IMLC), which can be used with the ALE-55 on a variety of aircraft.

AAR-57

Another major IDECM subsystem, the Lockheed Martin AAR-57 common missile warning system (C-MWS), has encountered problems in finding suitable locations for its six sensors on the F/A-18E/F. Ideally, the sensors should be located so they collectively provide full hemispherical coverage. But if installed in locations that can provide full coverage, the current AAR-57 sensors would compromise the aircraft's aerodynamic performance and degrade its low-observability characteristics. Boeing is working on a new sensor configuration that ideally can use the same internal subassemblies as a standard AAR-57 sensor. Because of the problem, the Navy decided that F/A-18E/Fs built under the initial LRIP program will not be outfitted with C-MWS, but hopes to install the system on later production aircraft. There are no plans to install C-MWS on F/A-18C/Ds.[243]

Armament

The F/A-18A/B/C/D is equipped with a single General Electric M61A1 20-mm cannon in the nose, and nine external stores stations—one at each wingtip, two under each wing, one on each fuselage corner just aft of the air intakes, and one under the fuselage centerline. The F/A-18E/F adds an additional station under each wing, for a total of 11. The stores stations are numbered sequentially from left to right. On A/B/C/Ds this means the left wingtip is No. 1, while the right wingtip is No. 9, with the fuselage being No. 5. Since there are two additional stations on the Super Hornet, the left wingtip is still No. 1, but the right wingtip is No. 11, with the fuselage centerline being No. 6.

An F/A-18A from VFMA-314 is displayed along with an impressive array of armaments. An AGM-88 HARM is on the wing pylon, while an AGM-84 Harpoon is in the foreground. A naval mine is off to the side. On the far side is an AGM-62 Walleye, a Mk 82 low-drag bomb, and a cluster bomb. *(Jeff Puzzullo via the Mick Roth Collection)*

Not to be outdone, the Canadians presented this CF-18B with its initial range of weapons. The M61 20-mm cannon is immediately in front of the aircraft, with 570 rounds of ammunition laid out in front of it. The two wheeled vehicles are used to reload the cannon. The 610-pound BL-755 cluster bombs are still in shipping containers. This is the standard cluster bomb used by the Australians, Canadians, and RAF, as well as many other countries. U.S. aircraft are also capable of carrying this weapon if the need arises. *(Tony Landis Collection)*

The wingtip stations were designed to carry AIM-9 Sidewinders, and can also carry various air combat maneuvering instrumentation pods when the aircraft is participating in exercises. The two air intake corner stations were designed to carry AIM-7 Sparrows, and have been cleared to carry AIM-120s. These stations are also used to carry various sensor pods when the Hornet is operating in the attack role. On the A/B/C/D, all wing stations can carry 2,600 pounds. On the Super Hornet, the two inboard stations on each wing can carry 2,600 pounds, while the outer wing station can only carry 1,150 pounds. On the Hornet, the inboard station on each wing and the fuselage centerline are plumbed to accept external fuel tanks. On the Super Hornet the inboard and "midboard" wing stations and fuselage centerline are plumbed for fuel tanks.

AN/AYQ-9(V)

The Smiths Industries AYQ-9(V) weapons control and management system provides fully computerized control of stores release to optimize accuracy and minimize crew workload. Although available for a variety of attack aircraft, the system is standard equipment on the F/A-18. Functions include maintaining an inventory of stores types, locations, quantity, status, and special conditions such as a hung store or locked pylon. Fifty types of weapons can be controlled by the system, including Sidewinder, Sparrow, AMRAAM, Maverick, Harpoon, SLAM, and HARM missiles; B61 nuclear weapons; several types of free-fall, retarded, and smart bombs; FLIR cameras; and the M61 cannon. Eleven types of mines can also be accommodated. The AYQ-9(V) system consists of a processor and control unit, together with decoders at each stores station plus the internal cannon. The system interfaces with the standard AYK-14 mission computer via a MIL-STD-1553 data bus.

The AYQ-9(V) provides for the operational readiness assessment of each store and weapons station, prepares the suspension and release equipment (including power-up, alignment, and fuzing), and activates the stores sequencing, arming, and release, including off-boresight guidance for air-to-air missiles. The F/A-18E/F uses an upgraded version of the system that accommodates up to 88 different weapons types (instead of 50) and is smaller, lighter, and easier to maintain.[244]

M61A1

A single General Electric Vulcan six-barrel 20-mm cannon with 578 rounds is mounted internally directly in front of the windscreen. The barrels are elevated approximately 2° to improve air-to-air target tracking. Each of the six barrels fires only once during each revolution of the barrel cluster, contributing to long weapon life by minimizing barrel erosion and heat generation.

The cannon has a selectable rate of fire of 4,000 or 6,000 rounds per minute and a muzzle velocity of 3,400 feet per second. The ingestion of gun gases into the engines is prevented by a fixed deflector that splits the muzzle blast and diverts gun gases to each side of the aircraft above the leading edge extension. There are three holes in the upper fuselage in front of the gun—one central hole for the rounds to pass through, and one on each side for gun gas ejection. Vents on the underside of the nose prevent the buildup of potentially dangerous gases in the gun bay.

The cannon is aimed with the HUD using the Boeing director gunsight, with a conventional gunsight as a backup. The designers took great measures to ensure that the cannon placement had no adverse effects (vibration, gun gas, etc.) on the radar components located near it. Conversely, the ammunition drum is well shielded against electromagnetic energy emanating from the radar.[245] The cannon, including its ammunition drum and feed mechanism, can be removed quickly by being winched down through the underside of the nose. On later F/A-18Ds the cannon and its associated gun doors can be replaced by a sensor pallet for reconnaissance missions.[246]

AIM-7

When the Hornet operates in the fighter role, AIM-7 Sparrow III semiactive radar homing air-to-air missiles are usually installed on LAU-116 ejectors on the intake stations. Hornets can carry as many as six Sparrows, one on each side of the fuselage and one on each wing station on LAU-115As. Super Hornets can carry eight AIM-7s, one on each side of the fuselage and one on each wing station.

This RAAF AF-18A from No. 77 Squadron fires an AIM-7M Sparrow while carrying two AGM-84 Harpoons, AIM-9 Sidewinders, and fuel tanks. The RAAF are currently integrating the AIM-120 into their Hornets, and the new missile will eventually replace the Sparrow as the standard beyond-visual-range weapon. They are also cooperating with the RAF to replace the Sidewinder with the new AIM-132 ASRAAM within-visual-range missile. *(RAAF/LAC Jason Freeman via Chris Hake)*

The Sparrow uses semiactive radar homing and is compatible with either pulse-Doppler or continuous wave (CW) radar illumination, although the Hornet does not have a CW capability. The AIM-7 is said to be effective out to distances of 25 miles, although the true effective range varies greatly with the conditions of the encounter. The AIM-7M version is 12 feet long and has a launch weight of about 500 pounds. The missile has two sets of delta-shaped fins—a set of fixed fins at the rear of the missile and a set of movable fins at the middle of the missile for steering. The 88-pound explosive warhead is contained in a stainless steel drum, which shatters upon detonation into 2,600 fragments, greatly increasing the prospect of a kill. The Sparrow can be detonated by impact or proximity fuses.

The AIM-7 was initially the Hornet's primary air-to-air weapon, but is has since been largely replaced by the AIM-120 AMRAAM. Although the Sparrow of today is a much more capable weapon than the Sparrow used in Vietnam, it still requires that the target be continually illuminated by the aircraft's radar transmitter in order for it to home in on reflected radar energy. This means that the Hornet can fire on only one target at a time with this weapon, which makes the fighter extremely vulnerable to attack by other enemy fighters.

AIM-9

The Sidewinder is an infrared-homing air-to-air missile. Hornets can carry two Sidewinders on each of the wing stations and one on each LAU-7 wingtip rail, for a total of 10 missiles. Super Hornets can also carry two on the two inner wing stations (⅜ and ⅝), one on the outer wing stations (2 and 10), and one on each wingtip rail, for a total of twelve missiles. The wingtip launch rails are inclined slightly downward, reflecting the Hornet's wing twist.

The Sidewinder is 9.4 feet long and has a wingspan of 25 inches and a diameter of 5 inches. The missile has four tail fins on the rear, with a "rolleron" at the tip of each fin that is spun at high speed by the slipstream in order to provide roll stability. The missile is steered by four canard fins mounted just behind the infrared seeker head. The Sidewinder has a launch weight of 180 pounds and a maximum effective range of about 10 miles. The blast-fragmentation warhead weighs 22 pounds and is detonated by either impact or proximity fusing.

AIM-9X

The Raytheon AIM-9X is the next-generation short-range air-to-air missile for the United States military. It consists of a Sidewinder rocket motor and warhead mated to new control surfaces and seeker. This is a compromise between total performance and cost. The goal is to produce a missile with better maneuverability than the AIM-9, with a seeker capable of acquiring targets at least 90° off boresight. The AIM-9X is designed to work with the JHMCS as part of the high off-boresight system (HOBS).

A successful launch from an F-15 on 1 September 1999 resulted in the intercept of a remotely piloted QF-4. The launch was the first look-down, shoot-down engagement evaluating the ability of the missile's infrared seeker. The AIM-9X has now completed four separation and control test launches, two from the F/A-18 and two from the F-15. "This success comes on the heels of the successful F/A-18 guided attack against an F-4," said Captain Dave Venlet, program manager for the Navy's Air-to-Air Missile Systems organization. "These missile shots keep us on the path toward FY00 production approval." The AIM-9X is a competitor to the AIM-132 ASRAAM in the world's markets.

AIM-120

The Raytheon (GM-Hughes) advanced medium-range air-to-air missile (AMRAAM)[247] is intended to combine the BVR performance of the Sparrow in an airframe that is not much larger than the Sidewinder. The largest tactical advantage of the AMRAAM is its launch-and-leave capability. The C/D can carry one AIM-120 on each fuselage corner, and two on

The newest air-to-air weapon to be cleared for use on the F/A-18 is the AIM-120 AMRAAM. This F/A-18C (BuNo 163706) has just "dropped" an AIM-120 from the fuselage station. Another missile can be seen on the wing station. The AMRAAM can either be "dropped" or rail-launched, depending upon what station it is carried on. Currently the AIM-120 is not cleared for flight on the wingtip stations. *(U.S. Navy via the Terry Panopalis Collection)*

each wing station, for a total of 10 AMRAAMs. Many A/B models have the same capabilities, depending upon their software configuration and a wiring upgrade. The Super Hornet can carry one AIM-120 on the fuselage corner, one on the outer wing stations (2 and 10), and two on each of the other four wing stations (⅜ and ⅝), for a total of 12 AMRAAMs.

It should be noted the Super Hornet load-out discussed here is the final capability expected to be cleared by the time of the first deployment. At the end of 1999 the Super Hornet is cleared to carry only AIM-9s on the wingtips, AIM-7s on the outer wing stations (2 and 10), and AIM-120s on the outer wing stations (2 and 10) and fuselage corners.

The AIM-120 is an active radar homing missile with its own tracking radar and inertial navigation system and a data link. Immediately before launch, the Hornet's fire control system downloads the target's position and the anticipated intercept location to the missile via the MIL-STD-1760 data bus. As the missile flies to the target, the fighter downlinks updated guidance information to the missile. At a certain distance to the target, the missile activates its own radar and tracks the target autonomously. This frees the Hornet from having to illuminate the target all the way to final intercept, allowing the fighter to maneuver for self-protection or to chase other targets. The AIM-120 was declared operational with the F/A-18 in September 1993.[248]

The AIM-120A was the initial production version. This has been superseded by the AIM-120B with a reprogrammable signal processor that allows new ECM threat information to be loaded in the field. The AIM-120C version has clipped wings to enable it to fit into the weapons bay on the F-22, although it has been tested on the F/A-18E/F.

The AMRAAM is 11.97 feet long and 7 inches in diameter and has a wingspan of 20.7 inches. The AMRAAM is considerably lighter than the Sparrow that it replaces, weighing about 350 pounds at launch. It carries a 48-pound high-explosive directed-fragmentation warhead. Maximum speed is about Mach 4, and the maximum range is 35 to 45 miles.

AIM-132

The Matra BAe Dynamics advanced short-range air-to-air missile (ASRAAM) has been ordered into production by the RAF and RAAF. The first deliveries of ASRAAM missiles to the RAAF under Project AIR 5400 were made in mid-April 1999 by Matra BAe Dynamics and its local partner British Aerospace Australia. The two dummy missiles, a unique variant designed for the RAAF, are physically representative of the operational version and will be used in flight trials to demonstrate safe carriage on RAAF AF-18 aircraft.[249]

ASRAAM is claimed by its manufacturers to be the most advanced within-visual-range (WVR) missile in the world. It has been developed to meet a United Kingdom requirement for a fifth-generation short-range missile to arm its Eurofighter Typhoon aircraft. The missile features a digital infrared seeker to detect and track enemy aircraft and a fast and agile airframe to ensure a high probability of intercept. ASRAAM will initially enter service with the Royal Air Force on Tornados and Harriers. Australia is the first export customer for the missile.

AGM-62

The Walleye is a guided glide bomb designed to be used primarily against targets such as fuel tanks, tunnels, bridges, radar sites, port facilities, and ammunition depots. The Walleye does not have a propulsion section and must rely on its ability to glide to the target after release from the aircraft. The Walleye I extended-range data link (ERDL) utilizes a tone data link system, while Walleye II uses a differential-phase-shift-keyed digital data link designed to prevent signal jamming. The data link permits the weapon to send a video image of the target until impact, allowing the aircraft to control the weapon in flight and to either retarget or redefine the target aim point. The controlling aircraft can be the launching aircraft or any aircraft equipped with an AWW-9B data link pod.

AGM-65

The Maverick is an air-to-surface, rocket-propelled guided missile designed primarily for the destruction of hard point targets such as bunkers. It is capable of day or night operations, has sufficient standoff range to permit avoidance of enemy defenses, and uses terminal homing guidance for a launch-and-leave capability. The AGM-65E uses a laser seeker while the AGM-65F uses an IIR seeker, and the seeker sections can be interchanged with no other alterations to the missile. The Navy AGM-65E/F differs from previous Air Force Maverick missiles by incorporating a heavier warhead and a dual-thrust rocket motor. The LAU-117/A(V)2/A is a single-rail guided missile launcher that provides the mechanical and electronic interface between the missile and launch aircraft. All C/D and E/F Hornets can launch the Maverick, and Australian and Canadian A/Bs were also equipped with the capability.

AGM-84

Hornets can be configured to carry two over-the-horizon Boeing (McDonnell Douglas) AGM-84D Harpoon antiship missiles. The Harpoon uses an active radar seeker to track its target and has a range of 70 nautical miles. Harpoon employs a low-level cruise profile, active radar guidance with counter-countermeasures, and terminal maneuvering to assure maximum weapon effectiveness. On the F/A-18, Harpoons are carried on either AERO-65 or AERO-7/A bomb racks, usually on the outer underwing stations (or middle wing station on the E/F).

AGM-84E

The AGM-84E takes the basic Harpoon missile, replaces the active radar seeker with the infrared seeker from the AGM-65D Maverick, and adds an integrated INS/GPS navigation unit and the data link from the Walleye glide bomb. The missile flies autonomously until it is in the vicinity of the target, when the seeker and data link are automatically activated and transmit an image of the target either to the launch aircraft or a "buddy" director aircraft. The operator verifies the target, selects a final aimpoint, and locks the seeker. SLAM has an advertised CEP of less than 10 feet at its 50 nautical mile maximum range.

AGM-84H

The SLAM ER (expanded response) is an all-weather, over-the horizon, precision-strike missile. The missile adds larger wings, an improved data link, a larger warhead, and a revised nose that reduces drag and RCS. The SLAM ER has a range of more than 150 nautical miles with the ability to precisely attack both land and sea targets. SLAM ER combines man-in-the-loop control with a highly precise inertial navigation system, jam-resistant GPS, and a hardened data link. Viewing the target in real time prior to impact allows target identification, reduces collateral damage, provides an immediate indication of mission success, and permits the pilot to select an alternative aimpoint if desired. The Navy plans to upgrade its entire inventory of 700 SLAMs into the SLAM ER configuration.

AGM-88

F/A-18s are capable of carrying four Raytheon (Texas Instruments) high-speed antiradiation missiles (HARMs) for "Iron Hand" suppression of enemy air defenses (SEAD) missions. The AGM-45 Shrike, the forerunner of the HARM, could also be carried by the Hornet, but has been retired from operational service.

The HARM has a terminal homing capability that provides a launch-and-leave capability. The AGM-88 has the capability of discriminating a single target from a number of emitters in the environment. The AGM-88B version has an improved guidance section that incorporates

The additional stores station on each wing of the Super Hornet will greatly increase its weapons-carrying flexibility. This E model is carrying two AIM-9s, two AGM-88 HARMs, two AGM-154 JSOWs, and two 1,000-pound Mk 83 bombs. Normally, an external fuel tank would probably be mounted on the fuselage centerline station and sensor pods would occupy the fuselage corner stations. This aircraft is not equipped with the four ALE-47 countermeasures dispensers under the intakes. *(Boeing)*

an improved tactical software and electronically reprogrammable memory. The LAU-118(V)1/A launcher provides the mechanical and electrical interface between the missile and aircraft. It is a single-rail launcher modified from the AERO-5B-1 series. A unique mechanical configuration prohibits installation of the HARM missile on an unmodified AERO-5 launcher.

The current version of the HARM is the AGM-88C Block IV, which has a more sensitive seeker, a faster processor, and more on-board memory than the original HARM. As of 1999 this version had not been released for export, and foreign customers were being provided older AGM-88As from U.S. inventory instead. The HARM seeker can be activated while the missile is on the pylon, and is a significant source of information on threats and targets during a SEAD mission. However, the missile cannot determine the range to the target, and, unfortunately, neither can the F/A-18. The only aircraft[250] that can launch in a "range-known" mode are the EA-6B and the F-16CJ when carrying the AN/ASQ-213 HARM Targeting System (HST) pod. The HST has been tested on the F/A-18D but has not been procured for use.

On 10 September 1998, a Navy F/A-18C successfully fired a Block V HARM, which used a new "home-on-jam" capability to lock on to a simulated jammer. Other enhancements include greater ability to attack the last known geographic location if a threat radar goes off the air and improved capability against advanced radar waveforms.[251]

AGM-154

The Raytheon (Texas Instruments) joint standoff weapon (JSOW) uses a GPS/INS system for midcourse navigation and imaging infrared (IIR) for terminal homing. The JSOW is just over 13 feet long and weighs between 1,000 and 1,500 pounds. Like the Walleye, it is an unpowered glide weapon. The JSOW has been cleared for use on all models of the F/A-18.

VFA-22 participated in the operational evaluations of the AGM-154 joint standoff weapon. The weapon uses a small set of "pop-out" wings on top of the fuselage to give it a range of between 17 and 46 miles, depending upon the release altitude. The weapon glides to its target and is not equipped with an engine. Several different warheads are available for the JSOW, depending upon the target to be attacked. The first operational JSOW was delivered on 10 June 1998 to VFA-81 aboard the USS *Dwight D. Eisenhower. (U.S. Navy)*

The JSOW provides a standoff range of 17 miles for a low-level release and 46 miles for a high-altitude release. The weapon can turn through 180° to engage off-boresite targets. Once programmed, as long as it is released at such an altitude and range that it can aerodynamically reach its target, it will autonomously calculate the flight path and profile to hit its target. If released at high speeds, JSOW will delay wing deployment to avoid penalizing its range by drag. The weapon can also be programmed to attack a target from a specific heading and to fly between multiple programmed waypoints. A typical profile is to glide in at an altitude of 200 feet, pop up close to the target and dive in to dispense its payload from several hundred feet.

The AGM-154A (Baseline JSOW) is intended for use against soft targets such as parked aircraft, vehicles, SAM sites, and mobile command posts. It carries 145 BLU-97A/B combined effects bomblets (CEBs) that are deployed as the JSOW dives from medium altitudes. These are the same CEBs that are used on the CBU-87 cluster bomb. Each CEB has a conical-shaped charge that can penetrate 5 to 7 inches of armor, a main charge that produces about 300 high-velocity fragments, and a zirconium sponge incendiary element.

The AGM-154B is a specialized antiarmor weapon that carries six BLU-108/B sensor-fuzed weapon (SFW) submunitions. These are the same SFWs that are used in the CBU-97 cluster bomb. Planned improvements to the SFW submunition will include a better IR sensor and a warhead that will produce a slug and a shrapnel pattern.

The AGM-154C variant is intended specifically to replace the Walleye glidebomb. The Navy-only AGM-154C will use a combination of an IIR terminal seeker and a AAW-9B or AAW-13 data link to achieve point target accuracy through aim point refinement and human-in-the-loop guidance.

The Navy began Operational Evaluation (OPEVAL) testing of the JSOW in February 1997, including five live weapon launches against representative real-world threat targets parked in an area of San Clemente Island, California. The targets consisted of an unrevetted MiG-23 aircraft, two ZSU-23-4 antiaircraft gun platforms, two Russian trucks, and a surrogate of a surface-to-air missile system. The first production JSOW was delivered on 10 June 1998, and was deployed by VFA-81 to war games at Fallon, Nevada, later that month. VFA-81 is the first squadron certified to carry the JSOW and was the first east coast squadron to deploy the weapon, aboard USS *Dwight D. Eisenhower* (CVN-71).[252]

AGM-158

The joint air-to-surface standoff missile (JASSM) is a low-observable launch-and-leave precision-guided weapon capable of operating in adverse weather. It will have a range of 100 nautical miles and carry a 1,000-pound-class unitary warhead. JASSM's midcourse guidance is provided by a GPS/INS protected by a new antijam GPS null-steering antenna system. In the terminal phase, JASSM is guided by an IIR seeker and a pattern-matching autonomous target recognition system. The first aircraft to carry JASSM will be the B-52H, F-16C/D, and F/A-18E/F.

In early 1998 Lockheed Martin was selected as the JASSM production contractor and began delivering preproduction units for continued testing. The first 87 operational missiles were ordered in FY00, and the total program is valued at approximately $3,000 million. Some problems relating to the F-16 have resulted in a minor redesign of the missile, and the program has been delayed by about 10 months. The cost of each production missile is estimated at $487,000.[253]

ADM-141

The Brunswick Defense tactical air-launched decoy (TALD) is a 400-pound glider used to confuse enemy air defenses. F/A-18s delivered large numbers of the ADM-141 decoys during the Gulf War to expose Iraqi antiair positions. TALDs proved to be an effective tactic in fooling enemy air defense radars, causing them to energize and thus become vulnerable to SEAD aircraft. The TALD can be twin- or triple-mounted on canted vertical ejection racks (CVERs) and triple ejector racks (TERs), and has a range of 68 nautical miles when launched from 35,000 feet. An Improved-TALD (I-TALD) has been developed that adds a small Teledyne CAE312 turbojet (from the canceled Tacit Rainbow antiradiation missile). This engine gives the decoy a range of 120 nautical miles, but more important, permits the decoy to be launched from as low as 500 feet.

MALD

The Teledyne Ryan miniature air-launched decoy (MALD) carries an electronic emitter that makes the 91-inch-long vehicle look like a fighter or bomber to a radar, while the flight-control computer replicates the flight characteristics of those aircraft. MALD is supposed to trick enemy air defenders into tracking the decoy, which would allow aircraft to launch AGM-88 HARMs to destroy SAM sites. The MALD is powered by a 50-lbf Sundstrand TJ-50 turbojet, and Teledyne Ryan has relied heavily on commercial hardware to achieve a $30,000 unit cost. The decoy has a range of more than 250 nautical miles, an endurance of more than 20 minutes, and can operate as high as 30,000 feet. The F-16 is the test aircraft, but the B-52H would likely be the first aircraft to use MALD operationally. The F-15, F/A-18, F-22, Joint Strike Fighter, and B-1B are other potential MALD users.[254]

The Navy is investigating installing a low-power jamming system on the MALD. Unlike the EA-6Bs which operate from standoff distances, MALD would be launched from a fighter outside a SAM's engagement envelope, fly close to the radar, and start jamming. The reduced range decreases exponentially the amount of power needed to achieve the same jamming effect of a standoff system.

Despite the major advances in precision-guided munitions, simple economics will usually dictate that the Mk 80 "iron" bomb will continue to be used for many missions. Here an AF-18A from No. 77 Squadron drops snake-eye retarded bombs while the wingman releases flares. Note the snakes have been dropped in an unretarded mode; the fins normally open 3 to 4 feet under the aircraft in retarded mode. *(RAAF/LAC Jason Freeman via Chris Hake)*

Mk 80

The Mk 80 series of "dumb" bombs was developed during the 1950s and is currently the United States' standard general-purpose bomb. Mk 82, Mk 83, and Mk 84 bombs weigh 500, 1,000, and 2,000 pounds, respectively. All Mk 80 series bombs are similar in construction; they are cylindrical in shape and equipped with conical fins or retarders for external high-speed carriage. The BLU-109 is a 2,000-pound hard-target penetrator version. The bombs are usually equipped with mechanical M904 (nose) and M905 (tail) fuzes or radar-proximity FMU-113 airburst fuzes.

   F/A-18s can carry two Mk 82 500-pound bombs loaded on CVERs on each wing station and the centerline. The wing stations on the Hornet and the inner two stations (3/9 and 4/8) on the Super Hornet can carry two Mk 83 1,000-pound bombs. The centerline can also carry a Mk 83, singly mounted on the parent rack. Mk 84s are also carried singly mounted on the parent rack (wing or centerline).

Laser-Guided Bombs

The Night Attack Hornet's AAS-38A/B laser designator enables the delivery of highly accurate laser-guided bombs. The ATFLIR will perform the same function on the Super Hornet. Kits add special nose sections (consisting of a semiactive laser homing seeker, guidance electronics, and control fins) and a set of folding aerodynamic surfaces to normal Mk 80 series bombs. Examples of these are the 610-pound Mk 82–based GBU-12, the 1,083-pound Mk 83–based GBU-16, the 2,081-pound Mk 84–based GBU-10, and the 2,081-pound

BLU-109/B–based GBU-24. Hornets can also use the AGM-123 Skipper, a powered version of the GBU-16.

GBU-29/32

The joint direct attack munition (JDAM) is a high-accuracy all-weather conventional bomb. JDAM is intended to upgrade the existing inventory of Mk 80 series bombs and the BLU-109 penetrator by integrating a GPS/INS guidance kit. No changes to the bomb casing, explosive fill, or fuzing mechanism are required. The tail kit may be fitted to any Mk 80 series bomb: The 250-pound (Mk 81) and 500-pound (Mk 82) bombs are designated GBU-29 and GBU-30, respectively, while the 1,000-pound (Mk 83) and 2,000-pound (Mk 84/BLU-109) versions are designated GBU-31 and GBU-32. The JDAM position information is continuously updated by aircraft avionics via the MIL-STD-1760 interface prior to release. Once released, the bomb's GPS/INS guides the bomb to its target with a delivery accuracy of approximately 45 feet when GPS is available and less than 100 feet when GPS is absent or jammed. In October 1995 Boeing was awarded a contract for the first 4,635 JDAM kits at an average unit cost of $18,000, less than half the original $40,000 estimate. The FY00 procurement of JDAM will total 6,195, and the JDAM production rate is expected to exceed 1,000 kits per month during the planned procurement of 87,496 JDAMs.

FFAR

Unguided folding-fin aircraft rocket (FFAR) pods are also part of the F/A-18 inventory. These include the LAU-61 nineteen-tube and LAU-68 seven-tube 2.75-inch versions. All use the standard air-to-ground strafing sight and may be singly or twin-mounted on the wing stations. None of these pods can be directly attached to the SUU-63 pylon, and therefore always require VERs and CVERs. Twin mounting is achieved through use of the LAU-115 adapter. Australian and Canadian Hornets also carry LAU-5003 pods, each armed with 19 Bristol Aerospace CRV-7 2.75-inch rockets.

Used mainly during FAC missions, the folding-fin unguided rocket is still a viable weapon in many circumstances. This No. 2 Squadron AF-18A fires a series of CVR-7 rockets, which is the standard Australian, Canadian, and RAF version of the FFAR. *(RAAF/LAC Jason Freeman via Chris Hake)*

A flight of otherwise unarmed CF-18As dispenses flares. Like all A/B models, the Canadian aircraft were built with ALE-39 dispensers, but will receive ALE-47 units as part of their fleetwide upgrade. *(Mike Reyno/Skytech Images)*

A 1,000-pound laser-guided bomb falls away from an F/A-18C during tests near Pax River. Note the cameras mounted under the nose, rear fuselage, and wingtips. The Hornet can carry two of these weapons, plus sensor pods, fuel tanks, and wingtip-mounted AIM-9s. This aircraft was assigned to the Strike Aircraft Test Directorate of the Naval Air Warfare Center Aircraft Division. *(U.S. Navy/DVIC DN-SC-95-01063)*

Nuclear Weapons

Although it is rarely talked about, U.S. Hornets can also carry two 10- to 100-kiloton B61 boosted-fission tactical nuclear weapons. The Hornet could also carry the 5- to 20-kiloton B57 fission weapon prior to its removal from the inventory. The B61 (mod 1 and 7) is a thermonuclear weapon with a yield of 10 to 500 kilotons. The B61-1 is 11.8 feet long and 13 inches in diameter and weighs 740 pounds. It is configured with four tail fins in an X configuration and uses either spin rockets or a parachute for stabilization. Of approximately 3,000 operational B61s delivered, only a few hundred mod 1 and mod 7 versions remain in service. The mod 7 is the most advanced version and uses a new electronic safety system and insensitive trigger explosive. The BDU-38/B is a full-size practice bomb that simulates the B61 nuclear weapon. It has the same dimensions, weight, and release modes as the actual B61 weapon. Needless to say, a nuclear strike is not a typically envisaged Hornet mission. The equipment necessary to arm and release nuclear weapons was not included on any export aircraft, and was removed from ex-U.S. aircraft sold to Spain prior to the transfer.

Cluster Bombs

All underwing stations can accommodate a variety of U.S. cluster bombs, as well as the 610-pound BL-755 cluster bomb unit (CBU) used by Australia and Canada.

The 500-pound CBU-59B Rockeye delivers 247 Mk 118 dart-shaped bomblets over an area of 3,300 square yards. The same dispenser allows the CBU-78 GATOR to spread 45 antitank and 15 antipersonnel mines.

The CBU-87 combined effects munition (CEM) consists of a SUU-65 tactical munitions dispenser (TMD) with an optional FZU-39/B proximity sensor. A total of 202 BLU-97/B CEBs are loaded in each dispenser, enabling an attack with wide-area coverage. The bomblet case is made of scored steel designed to break into approximately 300 preformed fragments for defeating light armor and personnel. There are 10 different height-of-burst selections that can be made before the weapon is dropped. The weapon is spin-stabilized during descent, with six different spin settings selectable before release. The weapon is 92

Space aboard carriers is limited. Here are four AIM-9 Sidewinders and two 500-pound bombs on one of the catwalks around the flight deck. (*U.S. Navy*)

The United States also uses unguided rockets, in this case 5-inch Zunis being launched from a pair of Marine F/A-18Cs. *(Tony Landis Collection)*

inches long and 15.6 inches in diameter and weighs 950 pounds. During FY90 the CBU-87 acquisition cost was $13,941 per weapon.

The CBU-89 Gator Mine is a 1,000-pound cluster munition consisting of a SUU-64/B TMD with 72 BLU-91/B antitank mines, 22 BLU-92/B antipersonnel mines, and an optional FZU-39/B proximity sensor. The BLU-91/B antitank mine contains microelectronics that can detect armored vehicles and detonate the mine when the target reaches the most vulnerable approach point. A Misznay-Schardin explosive charge defeats the belly armor of most vehicles. The BLU-92/B antipersonnel mines serve to discourage minefield clearing. Upon activation, the BLU-92/B explosion sends high-velocity fragments in a horizontal plane over a wide area.

Both mines have a programmable self-destruct feature that permits the battlefield commander to control the timing of a counterattack or defensive maneuver. The self-destruct time is set just prior to aircraft takeoff by using a simple selector switch on the dispenser, permitting a high degree of tactical flexibility during combat operations. After dispenser opening, the mines are self-dispersed by aerodynamic forces. The mine pattern on the ground is directly proportional to opening altitude, which is controlled by either the dispenser electromechanical fuze or an optional proximity sensor.

The CBU-97 sensor-fuzed weapon (SFW) consists of the SUU-66/B TMD, which opens and dispenses 10 parachute-stabilized BLU-108/B submunitions. Each of the 10 submunitions contains four 5-inch-diameter armor-penetrating "Skeet" projectiles with infrared sensors to detect armored targets. At a preset altitude sensed by a radar altimeter, a rocket motor fires to spin the submunition and initiate an ascent. The submunition then releases its four projectiles, which are lofted over the target area. When the Skeet detects a heat source such as a tank engine, it fires an explosively formed penetrator slug downward through the top of the target vehicle. Each Skeet scans 29,000 square feet; the 40 Skeets in each CBU-97 search a total of approximately 650,000 square feet. If no target is detected after a period of time, the projectile fires automatically after a preset time interval. The weapon was designed to be employed from altitudes between 200 and 20,000 feet at speeds between 250 and 650 knots. The CBU-97 is 92 inches long and 16 inches in diameter and weighs 927 pounds. The baseline cost in FY90 was $360,000 per CBU, although this has subsequently been reduced to $260,000.

Other Weapons

Among a host of others, Hornets can also use Mk 55 bottom mines, Mk 65 QuickStrike mines, and Mk 60 CAPTOR encapsulated Mk 46 torpedoes. The Mk 62 QuickStrike mine is an aircraft-laid bottom mine for use against submarines and surface targets, and is a conversion of the 500-pound Mk 82 general-purpose bomb. The Mk 65 is a 2,000-pound weapon, employing a thin-walled, mine-type case, as opposed to the thick-walled bomb-type case of the Mk 62. Other differences in the Mk 65 include a special arming device, a nose fairing, and a tail section adaptable to a parachute option. Mk 77 fuel gel canisters, or napalm, may be carried by the F/A-18, but are not authorized for carrier operations and are limited to shore-based flights.

Hornet Operators

U.S. Marine Corps

Squadron	Name	Base	From	To	Aircraft
VMFAT-101#	Sharpshooters	MCAS Miramar	1987	Present	F/A-18A/B/C/D
VMFA-112*	Cowboys	JRB Fort Worth	1991	Present	F/A-18A/B
VMFA-115	Silver Eagles	MCAS Beaufort	1985	Present	F/A-18A
VMFA(AW)-121	Green Knights	MCAS Miramar	1989	Present	F/A-18D(NA)
VMFA-122	Crusaders	MCAS Beaufort	1986	Present	F/A-18A
VMFA-134*	Smoke	MCAS Miramar	1989	Present	F/A-18A
VMFA-142*	Flying Gators	NAS Atlanta	1990	Present	F/A-18A
VMFA-212	Lancers	MCAS Miramar	1988	Present	F/A-18C
VMFA(AW)-224	Bengals	MCAS Beaufort		Present	F/A-18D(NA)
VMFA(AW)-225	Vikings	MCAS Miramar	1991	Present	F/A-18D(NA)
VMFA-232	Red Devils	MCAS Miramar	1989	Present	F/A-18C
VMFA-235	Death Angels	MCAS Miramar	1989	1996	F/A-18C
VMFA(AW)-242	Bats	MCAS Miramar	1991	Present	F/A-18D(NA)
VMFA-251	Thunderbolts	MCAS Beaufort	1986	Present	F/A-18C(NA)
VMFA-312	Checkerboards	MCAS Beaufort	1988	Present	F/A-18C(NA)
VMFA-314	Black Knights	MCAS Miramar	1982	Present	F/A-18C(NA)
VMFA-321*	Hell's Angels	Andrews AFB	1991	Present	F/A-18A/B
VMFA-323	Death Rattlers	MCAS Miramar	1982	Present	F/A-18C(NA)
VMFA(AW)-332	Moonlighters	MCAS Beaufort		Present	F/A-18D(NA)
VMFA-333	Shamrocks	MCAS Beaufort	1987	1992	F/A-18A

U.S. Marine Corps (Continued)

Squadron	Name	Base	From	To	Aircraft
VMFA-451	Warlords	MCAS Beaufort	1987	1997	F/A-18A
VMFA-531	Grey Ghosts	MCAS El Toro	1983	1992	F/A-18A
VMFA(AW)-533	Hawks	MCAS Beaufort		Present	F/A-18D(NA)

Normally, each Marine squadron is assigned 12 aircraft.
#fleet readiness squadron (training).
*reserve squadron.

U.S. Navy

Squadron	Name	Base	From	To	Aircraft
VAQ-34*	Electric Horsemen	NAS Lemoore		1993	F/A-18A/B
VF-45	Blackbirds	NAS Key West		1996	F/A-18A
VFA-15	Valions	NAS Oceana	1986	Present	F/A-18C(NA)
VFA-22	Fighting Redcocks	NAS Lemoore	1990	Present	F/A-18C(NA)
VFA-25	Fist of the Fleet	NAS Lemoore	1983	Present	F/A-18C(NA)
VFA-27	Royal Maces	NAF Atsugi	1991	Present	F/A-18C(NA)
VFA-34	Blue Blasters	NAS Oceana	1996	Present	F/A-18C(NA)
VFA-37	Bulls	NAS Oceana	1991	Present	F/A-18C(NA)
VFA-81	Sunliners	NAS Oceana	1988	Present	F/A-18C
VFA-82	Marauders	NAS Beaufort	1987	Present	F/A-18C
VFA-83	Rampagers	NAS Oceana	1988	Present	F/A-18C
VFA-86	Sidewinders	NAS Beaufort	1987	Present	F/A-18C
VFA-87	Golden Warriors	NAS Oceana	1986	Present	F/A-18C(NA)
VFA-94	Mighty Shrikes	NAS Lemoore	1991	Present	F/A-18C(NA)
VFA-97	Warhawks	NAS Lemoore	1991	Present	F/A-18A
VFA-105	Gunslingers	NAS Oceana	1991	Present	F/A-18C(NA)
VFA-106#	Gladiators	NAS Oceana	1984	Present	F/A-18A/B/C/D
VFA-113	Stingers	NAS Lemoore	1983	Present	F/A-18C(NA)
VFA-115	Eagles	NAS Lemoore	1996	Present	F/A-18C(NA)
VFA-122	Flying Eagles	NAS Lemoore	1999	Present	F/A-18E
VFA-125#	Rough Raiders	NAS Lemoore	1981	Present	F/A-18A/B/C/D
VFA-127	Desert Bogies	NAS Fallon	1992	1996	F/A-18A/B
VFA-131	Wildcats	NAS Oceana	1983	Present	F/A-18C(NA)
VFA-132	Privateers	NAS Cecil Field	1984	1992	F/A-18A
VFA-136	Knighthawks	NAS Oceana	1985	Present	F/A-18C(NA)
VFA-137	Kestrels	NAS Lemoore	1985	Present	F/A-18C(NA)

U.S. Navy *(Continued)*

Squadron	Name	Base	From	To	Aircraft
VFA-146	Blue Diamonds	NAS Lemoore	1989	Present	F/A-18C(NA)
VFA-147	Argonauts	NAS Lemoore	1989	Present	F/A-18C(NA)
VFA-151	Vigilantes	NAS Lemoore	1986	Present	F/A-18C(NA)
VFA-161	Chargers	NAS Lemoore	1986	1988	F/A-18A
VFA-192	Golden Dragons	NAF Atsugi	1985	Present	F/A-18C
VFA-195	Dambusters	NAF Atsugi	1985	Present	F/A-18C
VFA-203*	Blue Dolphins	NAS Atlanta	1989	Present	F/A-18A
VFA-204*	River Rattlers	JRB New Orleans	1991	Present	F/A-18A
VFA-303	Golden Hawks	NAS Lemoore	1984	1994	F/A-18A
VFA-305	Lobos	NAS Point Mugu	1987	1994	F/A-18A
VFC-12*	Fighting Omars	NAS Oceana	1988	Present	F/A-18A/B
VFC-13*	Saints	NAS Fallon	1988	1996	F/A-18A/B
VX-4	Evaluators	NAS Point Mugu	1980	1994	F/A-18A/C/D
VX-5	Vampires	NAWS China Lake		1994	F/A-18A/B/C/D
VX-9	Evaluators	NAWS China Lake	1994	Present	F/A-18C/D/E/F
NAWC-WD	Dust Devils	NAWS China Lake		Present	F/A-18C/D/E/F
NFDS	Blue Angels	NAS Pensacola	1987	Present	F/A-18A/B
NFWS	Top Gun	NAS Miramar		1996	F/A-18A/B
NSATS	Salty Dog†	NAS Patuxent River		Present	F/A-18C/D/E/F
NSAWC		NAS Fallon	1996	Present	F/A-18A/B
NSWC	Strike U	NAS Fallon		1996	F/A-18A/B
NWEF		Kirtland AFB			F/A-18A
USNTPS	Tea Kettle†	NAS Patuxent River	1986	Present	F/A-18B/D

Normally, each Navy squadron is assigned 12 aircraft.

*reserve squadron.

#fleet readiness squadron (training).

†This is actually a call sign, not a squadron name. This squadron does not have an official name.

NSAWC = Naval Strike Air Warfare Center; NSATS = Naval Strike Aircraft Test Squadron; NFWS = Navy Fighter Weapons School (combined into NSAWC); NSWC = Naval Strike Warfare Center (combined into NSAWC); NFDS = Navy Flight Demonstration Team.

VX-4 and VX-5 were combined into VX-9 in 1994.

Royal Australian Air Force

Squadron	Name	Base	From	To	Aircraft
No. 2 OCU #		RAAF Williamtown	1985	Present	AF-18A/B
No. 3 Sqn		RAAF Williamtown	1986	Present	AF-18A/B
No. 75 Sqn		RAAF Tindal	1988	Present	AF-18A/B

Royal Australian Air Force *(Continued)*

Squadron	Name	Base	From	To	Aircraft
No. 77 Sqn		RAAF Williamtown	1987	Present	AF-18A/B
ARDU		RAAF Edinburgh	1985	Present	AF-18A/B

Normally, each Australian squadron is assigned 18 aircraft.
#Operational Conversion Unit.
ARDU = Aircraft Research and Development Unit.

Canadian Forces

Squadron	Name	Base	From	To	Aircraft
No. 409 Sqn	Nighthawks	Baden-Soellingen	1984	1991	CF-18A/B
No. 410 TF(OT)S#	Cougars	CFB Cold Lake	1982	Present	CF-18A/B
No. 416 Sqn	Lynxes	CFB Cold Lake	1988	Present	CF-18A/B
No. 421 Sqn	Red Indians	Baden-Soellingen	1986	1992	CF-18A/B
No. 425 Sqn	Alouettes	CFB Bagotville	1985	Present	CF-18A/B
No. 433 Sqn	Porcupines	CFB Bagotville	1987	Present	CF-18A/B
No. 439 Sqn	Tigers	Baden-Soellingen	1987	1992	CF-18A/B
No. 441 Sqn	Silver Foxes	CFB Cold Lake	1985	Present	CF-18A/B
AETE		CFB Cold Lake	1982	Present	CF-18A/B

Traditionally, each Canadian squadron is assigned 18 aircraft, although the four operational CF-18 squadrons had only 15 each in 1999, while No. 410 Squadron had 25.
#Operational Conversion Unit.
AETE = Aerospace Engineering Test Establishment.

Ilmavoimat (Finnish Air Force)

Squadron	Name	Base	From	To	Aircraft
HävLLv 11	Lapin Wing	Rovaniemi	1998	Present	F-18C/D
HävLLv 21	Satakunnan Wing	Tampere-Pirkkala	1995	Present	F-18C/D
HävLLv 31	Karjalan Wing	Kuopio-Rissala	1996	Present	F-18C/D
Koelentokeskus		Halli	1997	Present	F-18C/D

Koelentokeskus = Flight Test Center.

Al Quwwat al Jawwiya al Kuwaitiya (Kuwait Air Force)

Squadron	Name	Base	From	To	Aircraft
No. 9 Sqn		Ahemd al Jaber AB	1993	Present	F/A-18C/D
No. 25 Sqn		Ali al Salem AB	1992	Present	F/A-18C/D

Tentara Udara Diraja Malaysia (Royal Malaysian Air Force)

Squadron	Name	Base	From	To	Aircraft
18 Night Fighter Skuadron		Butterworth	1997	Present	F/A-18D

Eército del Aire Español (Spanish Air Force)

Squadron	Name	Base	From	To	Aircraft
Escuadron 121		Torrejon de Ardoz	1986	Present	EF-18A/B+
Escuadron 122		Torrejon de Ardoz	1987	Present	EF-18A/B+
Escuadron 124#		Torrejon de Ardoz	1992	1994	EF-18A/B+
Escuadron 151		Zaragoza-Valenzuela	1989	Present	EF-18A/B+
Escuadron 152		Zaragoza-Valenzuela	1989	Present	EF-18A/B+
Escuadron 153#		Zaragoza-Valenzuela	1995	Present	EF-18B+
Escuadron 211		Morón	1996	Present	EF-18A/B+
CLAEX		Torrejón	1987	Present	EF-18A+

#Operational Conversion Unit.
CLAEX = *Centro Logistico de Armamento y Experimentación* = Weapons and Test Center.

Schweizerische Flugwaffe/Troupe d'Aviation Suisse (Swiss Air Force)

Squadron	Name	Base	From	To	Aircraft
Fliegerstaffel 11		Meiringen	1999	Present	F-18C/D
Fliegerstaffel 17		Payerne	1997	Present	F-18C/D
Fliegerstaffel 18		Sion	1998	Present	F-18C/D

Generally, aircraft in Swiss service are not assigned to a particular squadron on a permanent basis, but are part of a pool that distributes aircraft to the squadrons on the basis of need.

In addition to the bases listed above, Dubendorf has been equipped to operate F-18s.

Hornet Serial Numbers

This appendix begins with a Hornet Production Summary, pp. 186 and 187.

Hornet Production Summary

FY	Lot	Block	United States								Australia	
			F/A-18A	F/A-18B	F/A-18C	F/A-18D	F/A-18C(NA)	F/A-18D(NA)	F/A-18E	F/A-18F	AF-18A	AF-18B
76	I	1	3									
		2	3	1								
		3	3	1								
79	III	4	7	2								
80	IV	5	3	4								
		6	8	1								
		7	9									
81	V	8	10	4								
		9	17	4								
		10	23	3								
82	VI	11	17	3								
		12	20	1								
		13	22									
83	VII	14	18	3							3	7
		15	28	2							4	
		16	33								4	
84	VIII	17	24	3							7	
		18	25	4								5
		19	27	1							3	2
85	IX	20	24	3							6	
		21	26	1							5	
		22	30								4	2
86	X	23			23	8					4	2
		24			21	8					4	
		25			17	7					5	
87	XI	26			25	3					4	
		27			26	2					3	
		28			25	3					1	
88	XII	29					20	10				
		30					16	10				
		31					18	10				
89	XIII	32					17	7				
		33					23	7				
		34					24	6				
90	XIV	35					22					
		36					11	10				
		37					13	10				
91	XV	38					12	4				
		39					12	4				
		40					12	4				
92	XVI	41					9	9				
		42					12	3				
		43					13	2				
93	XVII	44					6	6				
		45					6	6				
		46					12					
94	XVIII	47					12		2			
		48					12		1	1		
		49					12		2	1		
95	XIX	50					24					
96	XX	51					10	8				
97		52					1	14	8	4		
98	XXI	53							8	12		
99	XXII	54							14	16		
00	XXIII	55							15	21		
01		56							14	28		
			380	41	137	31	330	130	64	83	57	18

Canada		Finland		Kuwait		Malaysia	Spain		Switzerland		
CF-188A	CF-188B	FN-18C	FN-18D	KAF-18C	KAF-18D	MAF-18D	EF-18A	EF-18B	SF-18C	SF-18D	Block
											1
											2
											3
											4
											5
											6
											7
	4										8
1	5										9
5	3										10
7											11
7	2										12
7	1										13
7	1										14
6	2										15
7	1										16
7	2							2			17
7	1						1	2			18
7	1							4			19
6	2						3	1			20
8							5	3			21
8							9				22
8							9				23
	9						6				24
	6						2				25
							5				26
							5				27
							7				28
							2				29
							4				30
							2				31
											32
											33
											34
				5	3						35
				3	3						36
				2	2						37
				8							38
				8							39
				6							40
											41
											42
											43
											44
											45
			4								46
		6	3								47
		2							8	6	48
		3							18	2	49
		11				8					50
		17									51
		18									52
											53
											54
											55
											56
98	40	57	7	32	8	8	60	12	26	8	

Hornet Serial Numbers

Type	Lot	Block	Contract Number	BuNo(s)	Quantity
			United States		
F/A-18A	I	1	N00019-75-C-0424	160775–160777	3
F/A-18A	I	2	N00019-75-C-0424	160778–160780	3
F/A-18B	I	2	N00019-75-C-0424	160781	1
F/A-18A	I	3	N00019-75-C-0424	160782–160783	2
F/A-18B	I	3	N00019-75-C-0424	160784	1
F/A-18A	I	3	N00019-75-C-0424	160785	1
F/A-18A	III	4	N00019-78-C-0526	161213–161216	4
F/A-18B	III	4	N00019-78-C-0526	161217	1
F/A-18A	III	4	N00019-78-C-0526	161248	1
F/A-18B	III	4	N00019-78-C-0526	161249	1
F/A-18A	III	4	N00019-78-C-0526	161250–161251	2
F/A-18A	IV	5	N00019-80-C-0173	161353	1
F/A-18B	IV	5	N00019-80-C-0173	161354–161357	4
F/A-18A	IV	5	N00019-80-C-0173	161358–161359	2
F/A-18B	IV	6	N00019-80-C-0173	161360	1
F/A-18A	IV	6	N00019-80-C-0173	161361–161367	7
F/A-18A	IV	6	N00019-80-C-0173	161519	1
F/A-18A	IV	7	N00019-80-C-0173	161520–161528	9
F/A-18A	V	8	N00019-81-C-0157	161702–161703	2
F/A-18B	V	8	N00019-81-C-0157	161704	1
F/A-18A	V	8	N00019-81-C-0157	161705–161706	2
F/A-18B	V	8	N00019-81-C-0157	161707	1
F/A-18A	V	8	N00019-81-C-0157	161708–161710	3
F/A-18B	V	8	N00019-81-C-0157	161711	1
F/A-18A	V	8	N00019-81-C-0157	161712–161713	2
F/A-18B	V	8	N00019-81-C-0157	161714	1
F/A-18A	V	8	N00019-81-C-0157	161715	1
F/A-18A	V	9	N00019-81-C-0157	161716–161718	3
F/A-18B	V	9	N00019-81-C-0157	161719	1
F/A-18A	V	9	N00019-81-C-0157	161720–161722	3
F/A-18B	V	9	N00019-81-C-0157	161723	1
F/A-18A	V	9	N00019-81-C-0157	161724–161726	3
F/A-18B	V	9	N00019-81-C-0157	161727	1

Hornet Serial Numbers *(Continued)*

Type	Lot	Block	Contract Number	BuNo(s)	Quantity
			United States		
F/A-18A	V	9	N00019-81-C-0157	161728–161732	5
F/A-18B	V	9	N00019-81-C-0157	161733	1
F/A-18A	V	9	N00019-81-C-0157	161734–161736	3
F/A-18A	V	10	N00019-81-C-0157	161737–161739	3
F/A-18B	V	10	N00019-81-C-0157	161740	1
F/A-18A	V	10	N00019-81-C-0157	161741–161745	5
F/A-18B	V	10	N00019-81-C-0157	161746	1
F/A-18A	V	10	N00019-81-C-0157	161747–161761	15
F/A-18B	V	10	N00019-81-C-0157	161924	1
F/A-18A	VI	11	N00019-82-0501	161925–161931	7
F/A-18B	VI	11	N00019-82-C-0501	161932	1
F/A-18A	VI	11	N00019-82-C-0501	161933–161937	5
F/A-18B	VI	11	N00019-82-C-0501	161938	1
F/A-18A	VI	11	N00019-82-C-0501	161939–161942	4
F/A-18B	VI	11	N00019-82-C-0501	161943	1
F/A-18A	VI	11	N00019-82-C-0501	161944	1
F/A-18A	VI	12	N00019-82-C-0501	161945–161946	2
F/A-18B	VI	12	N00019-82-C-0501	161947	1
F/A-18A	VI	12	N00019-82-C-0501	161948–161965	18
F/A-18A	VI	13	N00019-82-C-0501	161966–161987	22
F/A-18A	VII	14	N00019-83-C-0272	162394–162401	8
F/A-18B	VII	14	N00019-83-C-0272	162402	1
F/A-18A	VII	14	N00019-83-C-0272	162403–162407	5
F/A-18B	VII	14	N00019-83-C-0272	162408	1
F/A-18A	VII	14	N00019-83-C-0272	162409–162412	4
F/A-18B	VII	14	N00019-83-C-0272	162413	1
F/A-18A	VII	14	N00019-83-C-0272	162414	1
F/A-18A	VII	15	N00019-83-C-0272	162415–162418	4
F/A-18B	VII	15	N00019-83-C-0272	162419	1
F/A-18A	VII	15	N00019-83-C-0272	162420–162426	7
F/A-18B	VII	15	N00019-83-C-0272	162427	1
F/A-18A	VII	15	N00019-83-C-0272	162428–162444	17
F/A-18A	VII	16	N00019-83-C-0272	162445–162477	33

Hornet Serial Numbers *(Continued)*

Type	Lot	Block	Contract Number	BuNo(s)	Quantity
			United States		
F/A-18A	VIII	17	N00019-83-C-0272	162826–162835	10
F/A-18B	VIII	17	N00019-83-C-0272	162836	1
F/A-18A	VIII	17	N00019-83-C-0272	162837–162841	5
F/A-18B	VIII	17	N00019-83-C-0272	162842	1
F/A-18A	VIII	17	N00019-83-C-0272	162843–162849	7
F/A-18B	VIII	17	N00019-83-C-0272	162850	1
F/A-18A	VIII	17	N00019-83-C-0272	162851–162852	2
F/A-18A	VIII	18	N00019-83-C-0272	162853–162856	4
F/A-18B	VIII	18	N00019-83-C-0272	162857	1
F/A-18A	VIII	18	N00019-83-C-0272	162858–162863	6
F/A-18B	VIII	18	N00019-83-C-0272	162864	1
F/A-18A	VIII	18	N00019-83-C-0272	162865–162869	5
F/A-18B	VIII	18	N00019-83-C-0272	162870	1
F/A-18A	VIII	18	N00019-83-C-0272	162871–162875	5
F/A-18B	VIII	18	N00019-83-C-0272	162876	1
F/A-18A	VIII	18	N00019-83-C-0272	162877–162881	5
F/A-18A	VIII	19	N00019-83-C-0272	162882–162884	3
F/A-18B	VIII	19	N00019-83-C-0272	162885	1
F/A-18A	VIII	19	N00019-83-C-0272	162886–162909	24
F/A-18A	IX	20	N00019-84-C-0063	163092–163103	12
F/A-18B	IX	20	N00019-84-C-0063	163104	1
F/A-18A	IX	20	N00019-84-C-0063	163105–163109	5
F/A-18B	IX	20	N00019-84-C-0063	163110	1
F/A-18A	IX	20	N00019-84-C-0063	163111–163114	4
F/A-18B	IX	20	N00019-84-C-0063	163115	1
F/A-18A	IX	20	N00019-84-C-0063	163116–163118	3
F/A-18A	IX	21	N00019-84-C-0063	163119–163122	4
F/A-18B	IX	21	N00019-84-C-0063	163123	1
F/A-18A	IX	21	N00019-84-C-0063	163124–163145	22
F/A-18A	IX	22	N00019-84-C-0063	163146–163175	30
F/A-18C	X	23	N00019-84-C-0270	163427–163433	7
F/A-18D	X	23	N00019-84-C-0270	163434	1
F/A-18C	X	23	N00019-84-C-0270	163435	1

Hornet Serial Numbers *(Continued)*

Type	Lot	Block	Contract Number	BuNo(s)	Quantity
			United States		
F/A-18D	X	23	N00019-84-C-0270	163436	1
F/A-18C	X	23	N00019-84-C-0270	163437–163440	4
F/A-18D	X	23	N00019-84-C-0270	163441	1
F/A-18C	X	23	N00019-84-C-0270	163442–163444	3
F/A-18D	X	23	N00019-84-C-0270	163445	1
F/A-18C	X	23	N00019-84-C-0270	163446	1
F/A-18D	X	23	N00019-84-C-0270	163447	1
F/A-18C	X	23	N00019-84-C-0270	163448–163451	4
F/A-18D	X	23	N00019-84-C-0270	163452	1
F/A-18C	X	23	N00019-84-C-0270	163453	1
F/A-18D	X	23	N00019-84-C-0270	163454	1
F/A-18C	X	23	N00019-84-C-0270	163455–163456	2
F/A-18D	X	23	N00019-84-C-0270	163457	1
F/A-18C	X	24	N00019-84-C-0270	163458–163459	2
F/A-18D	X	24	N00019-84-C-0270	163460	1
F/A-18C	X	24	N00019-84-C-0270	163461–163463	3
F/A-18D	X	24	N00019-84-C-0270	163464	1
F/A-18C	X	24	N00019-84-C-0270	163465–163467	3
F/A-18D	X	24	N00019-84-C-0270	163468	1
F/A-18C	X	24	N00019-84-C-0270	163469–163471	3
F/A-18D	X	24	N00019-84-C-0270	163472	1
F/A-18C	X	24	N00019-84-C-0270	163473	1
F/A-18D	X	24	N00019-84-C-0270	163474	1
F/A-18C	X	24	N00019-84-C-0270	163475–163478	4
F/A-18D	X	24	N00019-84-C-0270	163479	1
F/A-18C	X	24	N00019-84-C-0270	163480–163481	2
F/A-18D	X	24	N00019-84-C-0270	163482	1
F/A-18C	X	24	N00019-84-C-0270	163483–163485	3
F/A-18D	X	24	N00019-84-C-0270	163486	1
F/A-18C	X	25	N00019-84-C-0270	163487	1
F/A-18D	X	25	N00019-84-C-0270	163488	1
F/A-18C	X	25	N00019-84-C-0270	163489–163491	3
F/A-18D	X	25	N00019-84-C-0270	163492	1

Hornet Serial Numbers *(Continued)*

Type	Lot	Block	Contract Number	BuNo(s)	Quantity
			United States		
F/A-18C	X	25	N00019-84-C-0270	163493–163496	4
F/A-18D	X	25	N00019-84-C-0270	163497	1
F/A-18C	X	25	N00019-84-C-0270	163498–163499	2
F/A-18D	X	25	N00019-84-C-0270	163500–163501	2
F/A-18C	X	25	N00019-84-C-0270	163502–163506	5
F/A-18D	X	25	N00019-84-C-0270	163507	1
F/A-18C	X	25	N00019-84-C-0270	163508–163509	2
F/A-18D	X	25	N00019-84-C-0270	163510	1
F/A-18C	XI	26	N00019-85-C-0250	163699	1
F/A-18D	XI	26	N00019-85-C-0250	163700	1
F/A-18C	XI	26	N00019-85-C-0250	163701–163706	6
F/A-18D	XI	26	N00019-85-C-0250	163707	1
F/A-18C	XI	26	N00019-85-C-0250	163708–163719	12
F/A-18D	XI	26	N00019-85-C-0250	163720	1
F/A-18C	XI	26	N00019-85-C-0250	163721–163726	6
F/A-18C	XI	27	N00019-85-C-0250	163727–163733	7
F/A-18D	XI	27	N00019-85-C-0250	163734	1
F/A-18C	XI	27	N00019-85-C-0250	163735–163748	14
F/A-18D	XI	27	N00019-85-C-0250	163749	1
F/A-18C	XI	27	N00019-85-C-0250	163750–163754	5
F/A-18C	XI	28	N00019-85-C-0250	163755–163762	8
F/A-18D	XI	28	N00019-85-C-0250	163763	1
F/A-18C	XI	28	N00019-85-C-0250	163764–163770	7
F/A-18D	XI	28	N00019-85-C-0250	163771	1
F/A-18C	XI	28	N00019-85-C-0250	163772–163777	6
F/A-18D	XI	28	N00019-85-C-0250	163778	1
F/A-18C	XI	28	N00019-85-C-0250	163779–163782	4
F/A-18C	XII	29	N00019-86-C-0207	163985	1
F/A-18D	XII	29	N00019-86-C-0207	163986	1
F/A-18C	XII	29	N00019-86-C-0207	163987–163988	2
F/A-18D	XII	29	N00019-86-C-0207	163989	1
F/A-18C	XII	29	N00019-86-C-0207	163990	1
F/A-18D	XII	29	N00019-86-C-0207	163991	1

Hornet Serial Numbers *(Continued)*

Type	Lot	Block	Contract Number	BuNo(s)	Quantity
			United States		
F/A-18C	XII	29	N00019-86-C-0207	163992–163993	2
F/A-18D	XII	29	N00019-86-C-0207	163994	1
F/A-18C	XII	29	N00019-86-C-0207	163995–163996	2
F/A-18D	XII	29	N00019-86-C-0207	163997	1
F/A-18C	XII	29	N00019-86-C-0207	163998–163999	2
F/A-18C	XII	29	N00019-86-C-0207	164000	1
F/A-18D	XII	29	N00019-86-C-0207	164001	1
F/A-18C	XII	29	N00019-86-C-0207	164002–164004	3
F/A-18D	XII	29	N00019-86-C-0207	164005	1
F/A-18C	XII	29	N00019-86-C-0207	164006–164008	3
F/A-18D	XII	29	N00019-86-C-0207	164009	1
F/A-18C	XII	29	N00019-86-C-0207	164010	1
F/A-18D	XII	29	N00019-86-C-0207	164011	1
F/A-18C	XII	29	N00019-86-C-0207	164012–164013	2
F/A-18D	XII	29	N00019-86-C-0207	164014	1
F/A-18C	XII	30	N00019-86-C-0207	164015–164016	2
F/A-18D	XII	30	N00019-86-C-0207	164017	1
F/A-18C	XII	30	N00019-86-C-0207	164018	1
F/A-18D	XII	30	N00019-86-C-0207	164019	1
F/A-18C	XII	30	N00019-86-C-0207	164020–164021	2
F/A-18D	XII	30	N00019-86-C-0207	164022	1
F/A-18C	XII	30	N00019-86-C-0207	164023	1
F/A-18D	XII	30	N00019-86-C-0207	164024	1
F/A-18C	XII	30	N00019-86-C-0207	164025	1
F/A-18D	XII	30	N00019-86-C-0207	164026	1
F/A-18C	XII	30	N00019-86-C-0207	164027	1
F/A-18D	XII	30	N00019-86-C-0207	164028	1
F/A-18C	XII	30	N00019-86-C-0207	164029–164031	3
F/A-18D	XII	30	N00019-86-C-0207	164032	1
F/A-18C	XII	30	N00019-86-C-0207	164033–164034	2
F/A-18D	XII	30	N00019-86-C-0207	164035	1
F/A-18C	XII	30	N00019-86-C-0207	164036–164037	2
F/A-18D	XII	30	N00019-86-C-0207	164038	1

Hornet Serial Numbers *(Continued)*

Type	Lot	Block	Contract Number	BuNo(s)	Quantity
			United States		
F/A-18C	XII	30	N00019-86-C-0207	164039	1
F/A-18D	XII	30	N00019-86-C-0207	164040	1
F/A-18C	XII	31	N00019-86-C-0207	164041–164042	2
F/A-18D	XII	31	N00019-86-C-0207	164043	1
F/A-18C	XII	31	N00019-86-C-0207	164044–164045	2
F/A-18D	XII	31	N00019-86-C-0207	164046	1
F/A-18C	XII	31	N00019-86-C-0207	164047–164048	2
F/A-18D	XII	31	N00019-86-C-0207	164049	1
F/A-18C	XII	31	N00019-86-C-0207	164050	1
F/A-18D	XII	31	N00019-86-C-0207	164051	1
F/A-18C	XII	31	N00019-86-C-0207	164052	1
F/A-18D	XII	31	N00019-86-C-0207	164053	1
F/A-18C	XII	31	N00019-86-C-0207	164054–164055	2
F/A-18D	XII	31	N00019-86-C-0207	164056	1
F/A-18C	XII	31	N00019-86-C-0207	164057	1
F/A-18D	XII	31	N00019-86-C-0207	164058	1
F/A-18C	XII	31	N00019-86-C-0207	164059–164060	2
F/A-18D	XII	31	N00019-86-C-0207	164061	1
F/A-18C	XII	31	N00019-86-C-0207	164062–164063	2
F/A-18D	XII	31	N00019-86-C-0207	164064	1
F/A-18C	XII	31	N00019-86-C-0207	164065–164067	3
F/A-18D	XII	31	N00019-86-C-0207	164068	1
F/A-18D	XIII	32	N00019-88-C-0069	164196	1
F/A-18C	XIII	32	N00019-88-C-0069	164197	1
F/A-18D	XIII	32	N00019-88-C-0069	164198	1
F/A-18C	XIII	32	N00019-88-C-0069	164199–164202	4
F/A-18D	XIII	32	N00019-88-C-0069	164203	1
F/A-18C	XIII	32	N00019-88-C-0069	164204–164206	3
F/A-18D	XIII	32	N00019-88-C-0069	164207	1
F/A-18C	XIII	32	N00019-88-C-0069	164208–164210	3
F/A-18D	XIII	32	N00019-88-C-0069	164211	1
F/A-18C	XIII	32	N00019-88-C-0069	164212–164215	4
F/A-18D	XIII	32	N00019-88-C-0069	164216	1

Hornet Serial Numbers *(Continued)*

Type	Lot	Block	Contract Number	BuNo(s)	Quantity
			United States		
F/A-18C	XIII	32	N00019-88-C-0069	164217–164218	2
F/A-18D	XIII	32	N00019-88-C-0069	164219	1
F/A-18C	XIII	33	N00019-88-C-0069	164220–164223	4
F/A-18D	XIII	33	N00019-88-C-0069	164224	1
F/A-18C	XIII	33	N00019-88-C-0069	164225–164227	3
F/A-18D	XIII	33	N00019-88-C-0069	164228	1
F/A-18C	XIII	33	N00019-88-C-0069	164229–164232	4
F/A-18D	XIII	33	N00019-88-C-0069	164233	1
F/A-18C	XIII	33	N00019-88-C-0069	164234–164236	3
F/A-18D	XIII	33	N00019-88-C-0069	164237	1
F/A-18C	XIII	33	N00019-88-C-0069	164238–164240	3
F/A-18D	XIII	33	N00019-88-C-0069	164241	1
F/A-18C	XIII	33	N00019-88-C-0069	164242–164244	3
F/A-18D	XIII	33	N00019-88-C-0069	164245	1
F/A-18C	XIII	33	N00019-88-C-0069	164246–164248	3
F/A-18D	XIII	33	N00019-88-C-0069	164249	1
F/A-18C	XIII	34	N00019-88-C-0069	164250–164253	4
F/A-18D	XIII	34	N00019-88-C-0069	164254	1
F/A-18C	XIII	34	N00019-88-C-0069	164255–164258	4
F/A-18D	XIII	34	N00019-88-C-0069	164259	1
F/A-18C	XIII	34	N00019-88-C-0069	164260–164262	3
F/A-18D	XIII	34	N00019-88-C-0069	164263	1
F/A-18C	XIII	34	N00019-88-C-0069	164264–164266	3
F/A-18D	XIII	34	N00019-88-C-0069	164267	1
F/A-18C	XIII	34	N00019-88-C-0069	164268–164271	4
F/A-18D	XIII	34	N00019-88-C-0069	164272	1
F/A-18C	XIII	34	N00019-88-C-0069	164273–164278	6
F/A-18D	XIII	34	N00019-88-C-0069	164279	1
F/A-18C	XIV	35	N00019-88-C-0289	164627–164648	22
F/A-18D	XIV	36	N00019-88-C-0289	164649–164653	5
F/A-18C	XIV	36	N00019-88-C-0289	164654–164655	2
F/A-18D	XIV	36	N00019-88-C-0289	164656	1
F/A-18C	XIV	36	N00019-88-C-0289	164657–164658	2

Hornet Serial Numbers *(Continued)*

Type	Lot	Block	Contract Number	BuNo(s)	Quantity
			United States		
F/A-18D	XIV	36	N00019-88-C-0289	164659	1
F/A-18C	XIV	36	N00019-88-C-0289	164660–164661	2
F/A-18D	XIV	36	N00019-88-C-0289	164662	1
F/A-18C	XIV	36	N00019-88-C-0289	164663–164664	2
F/A-18D	XIV	36	N00019-88-C-0289	164665	1
F/A-18C	XIV	36	N00019-88-C-0289	164666	1
F/A-18D	XIV	36	N00019-88-C-0289	164667	1
F/A-18C	XIV	36	N00019-88-C-0289	164668–164669	2
F/A-18D	XIV	37	N00019-88-C-0289	164670	1
F/A-18C	XIV	37	N00019-88-C-0289	164671	1
F/A-18D	XIV	37	N00019-88-C-0289	164672	1
F/A-18C	XIV	37	N00019-88-C-0289	164673	1
F/A-18D	XIV	37	N00019-88-C-0289	164674	1
F/A-18C	XIV	37	N00019-88-C-0289	164675–164676	2
F/A-18D	XIV	37	N00019-88-C-0289	164677	1
F/A-18C	XIV	37	N00019-88-C-0289	164678	1
F/A-18D	XIV	37	N00019-88-C-0289	164679	1
F/A-18C	XIV	37	N00019-88-C-0289	164680–164682	3
F/A-18D	XIV	37	N00019-88-C-0289	164683	1
F/A-18C	XIV	37	N00019-88-C-0289	164684	1
F/A-18D	XIV	37	N00019-88-C-0289	164685	1
F/A-18C	XIV	37	N00019-88-C-0289	164686–164687	2
F/A-18D	XIV	37	N00019-88-C-0289	164688	1
F/A-18C	XIV	37	N00019-88-C-0289	164689	1
F/A-18D	XIV	37	N00019-88-C-0289	164690	1
F/A-18C	XIV	37	N00019-88-C-0289	164691	1
F/A-18D	XIV	37	N00019-88-C-0289	164692	1
F/A-18C	XV	38	N00019-90-C-0010	164693	1
F/A-18D	XV	38	N00019-90-C-0010	164694	1
F/A-18C	XV	38	N00019-90-C-0010	164695–164698	4
F/A-18D	XV	38	N00019-90-C-0010	164699	1
F/A-18C	XV	38	N00019-90-C-0010	164700–164701	2
F/A-18D	XV	38	N00019-90-C-0010	164702	1

Hornet Serial Numbers *(Continued)*

Type	Lot	Block	Contract Number	BuNo(s)	Quantity
			United States		
F/A-18C	XV	38	N00019-90-C-0010	164703–164704	2
F/A-18D	XV	38	N00019-90-C-0010	164705	1
F/A-18C	XV	38	N00019-90-C-0010	164706–164710	5
F/A-18D	XV	39	N00019-90-C-0010	164711	1
F/A-18C	XV	39	N00019-90-C-0010	164712–164713	2
F/A-18D	XV	39	N00019-90-C-0010	164714	1
F/A-18C	XV	39	N00019-90-C-0010	164715–164716	2
F/A-18D	XV	39	N00019-90-C-0010	164717	1
F/A-18C	XV	39	N00019-90-C-0010	164718–164722	5
F/A-18D	XV	39	N00019-90-C-0010	164723	1
F/A-18C	XV	39	N00019-90-C-0010	164724	1
F/A-18C	XV	40	N00019-90-C-0010	164725	1
F/A-18D	XV	40	N00019-90-C-0010	164726	1
F/A-18C	XV	40	N00019-90-C-0010	164727–164728	2
F/A-18D	XV	40	N00019-90-C-0010	164729	1
F/A-18C	XV	40	N00019-90-C-0010	164730–164734	5
F/A-18D	XV	40	N00019-90-C-0010	164735	1
F/A-18C	XV	40	N00019-90-C-0010	164736–164737	2
F/A-18D	XV	40	N00019-90-C-0010	164738	1
F/A-18C	XV	40	N00019-90-C-0010	164739–164740	2
F/A-18C	XVI	41	N00019-90-C-0285	164865	1
F/A-18D	XVI	41	N00019-90-C-0285	164866	1
F/A-18C	XVI	41	N00019-90-C-0285	164867	1
F/A-18D	XVI	41	N00019-90-C-0285	164868	1
F/A-18C	XVI	41	N00019-90-C-0285	164869	1
F/A-18D	XVI	41	N00019-90-C-0285	164870	1
F/A-18C	XVI	41	N00019-90-C-0285	164871	1
F/A-18D	XVI	41	N00019-90-C-0285	164872	1
F/A-18C	XVI	41	N00019-90-C-0285	164873	1
F/A-18D	XVI	41	N00019-90-C-0285	164874	1
F/A-18C	XVI	41	N00019-90-C-0285	164875	1
F/A-18D	XVI	41	N00019-90-C-0285	164876	1
F/A-18C	XVI	41	N00019-90-C-0285	164877	1

Hornet Serial Numbers *(Continued)*

Type	Lot	Block	Contract Number	BuNo(s)	Quantity
			United States		
F/A-18D	XVI	41	N00019-90-C-0285	164878	1
F/A-18C	XVI	41	N00019-90-C-0285	164879	1
F/A-18D	XVI	41	N00019-90-C-0285	164880	1
F/A-18C	XVI	41	N00019-90-C-0285	164881	1
F/A-18D	XVI	41	N00019-90-C-0285	164882	1
F/A-18C	XVI	42	N00019-90-C-0285	164883	1
F/A-18D	XVI	42	N00019-90-C-0285	164884	1
F/A-18C	XVI	42	N00019-90-C-0285	164885	1
F/A-18D	XVI	42	N00019-90-C-0285	164886	1
F/A-18C	XVI	42	N00019-90-C-0285	164887	1
F/A-18D	XVI	42	N00019-90-C-0285	164888	1
F/A-18C	XVI	42	N00019-90-C-0285	164889–164897	9
F/A-18D	XVI	43	N00019-90-C-0285	164898	1
F/A-18C	XVI	43	N00019-90-C-0285	164899–164900	2
F/A-18D	XVI	43	N00019-90-C-0285	164901	1
F/A-18C	XVI	43	N00019-90-C-0285	164902–164912	11
F/A-18D	XVII	44	N00019-92-C-0006	164945	1
F/A-18C	XVII	44	N00019-92-C-0006	164946	1
F/A-18D	XVII	44	N00019-92-C-0006	164947	1
F/A-18C	XVII	44	N00019-92-C-0006	164948	1
F/A-18D	XVII	44	N00019-92-C-0006	164949	1
F/A-18C	XVII	44	N00019-92-C-0006	164950	1
F/A-18D	XVII	44	N00019-92-C-0006	164951	1
F/A-18C	XVII	44	N00019-92-C-0006	164952	1
F/A-18D	XVII	44	N00019-92-C-0006	164953	1
F/A-18C	XVII	44	N00019-92-C-0006	164954	1
F/A-18D	XVII	44	N00019-92-C-0006	164955	1
F/A-18C	XVII	44	N00019-92-C-0006	164956	1
F/A-18D	XVII	45	N00019-92-C-0006	164957	1
F/A-18C	XVII	45	N00019-92-C-0006	164958	1
F/A-18D	XVII	45	N00019-92-C-0006	164959	1
F/A-18C	XVII	45	N00019-92-C-0006	164960	1
F/A-18D	XVII	45	N00019-92-C-0006	164961	1

Hornet Serial Numbers *(Continued)*

Type	Lot	Block	Contract Number	BuNo(s)	Quantity
			United States		
F/A-18C	XVII	45	N00019-92-C-0006	164962	1
F/A-18D	XVII	45	N00019-92-C-0006	164963	1
F/A-18C	XVII	45	N00019-92-C-0006	164964	1
F/A-18D	XVII	45	N00019-92-C-0006	164965	1
F/A-18C	XVII	45	N00019-92-C-0006	164966	1
F/A-18D	XVII	45	N00019-92-C-0006	164967	1
F/A-18C	XVII	45	N00019-92-C-0006	164968	1
F/A-18C	XVII	46	N00019-92-C-0006	164969–164980	12
F/A-18C	XVIII	47	N00019-93-C-0033	165171–165182	12
F/A-18C	XVIII	48	N00019-93-C-0033	165183–165194	12
F/A-18C	XVIII	49	N00019-93-C-0033	165195–165206	12
F/A-18C	XIX	50	N00019-94-C-0084	165207–165230	24
F/A-18C	XX	51	N00019-95-C-0015	165399–165408	10
F/A-18D	XX	51	N00019-95-C-0015	165409–165416	8
F/A-18C	XXI	52	N00019-97-C-0099	165526	1
F/A-18D	XXI	52	N00019-97-C-0099	165527–165532	6
F/A-18D	XXI	52	N00019-96-C-0146	165680–165687	8
F/A-18E	XVIII	47	N00019-92-C-0059	165164–165165	2
F/A-18F	XVIII	48	N00019-92-C-0059	165166	1
F/A-18E	XVIII	48	N00019-92-C-0059	165167	1
F/A-18E	XVIII	49	N00019-92-C-0059	165168–165169	2
F/A-18F	XVIII	49	N00019-93-C-0059	165170	1
F/A-18E	XXI	52	N00019-96-C-0065	165533–165540	8
F/A-18F	XXI	52	N00019-96-C-0065	165541–165544	4
F/A-18E	XXII	53	N00019-97-C-0136	165660–165667	8
F/A-18F	XXII	53	N00019-97-C-0136	165668–165679	12
F/A-18E	XXIII	54	N00019-97-C-0136	165779-165792	14
F/A-18F	XX	54	N00019-97-C-0136	165793-165808	16
			Canada		
CF-188B	V	8	Canada 2RF80-000472	188901–188904	4
CF-188A	V	9	Canada 2RF80-000472	188701	1
CF-188B	V	9	Canada 2RF80-000472	188905–188909	5
CF-188A	V	10	Canada 2RF80-000472	188702–188706	5

Hornet Serial Numbers *(Continued)*

Type	Lot	Block	Contract Number	BuNo(s)	Quantity
			Canada		
CF-188B	V	10	Canada 2RF80-000472	188910–188912	3
CF-188A	VI	11	Canada 2RF80-000472	188707–188713	7
CF-188A	VI	12	Canada 2RF80-000472	188714–188720	7
CF-188B	VI	12	Canada 2RF80-000472	188913–188914	2
CF-188A	VI	13	Canada 2RF80-000472	188721–188727	7
CF-188B	VI	13	Canada 2RF80-000472	188915	1
CF-188A	VII	14	Canada 2RF80-000472	188728–188734	7
CF-188B	VII	14	Canada 2RF80-000472	188916	1
CF-188A	VII	15	Canada 2RF80-000472	188735–188740	6
CF-188B	VII	15	Canada 2RF80-000472	188917–188918	2
CF-188A	VII	16	Canada 2RF80-000472	188741–188747	7
CF-188B	VII	16	Canada 2RF80-000472	188919	1
CF-188A	VIII	17	Canada 2RF80-000472	188748–188754	7
CF-188B	VIII	17	Canada 2RF80-000472	188920–188921	2
CF-188A	VIII	18	Canada 2RF80-000472	188755–188761	7
CF-188B	VIII	18	Canada 2RF80-000472	188922	1
CF-188A	VIII	19	Canada 2RF80-000472	188762–188768	7
CF-188B	VIII	19	Canada 2RF80-000472	188923	1
CF-188A	IX	20	Canada 2RF80-000472	188769–188774	6
CF-188B	IX	20	Canada 2RF80-000472	188924–188925	2
CF-188A	IX	21	Canada 2RF80-000472	188775–188782	8
CF-188A	IX	22	Canada 2RF80-000472	188783–188790	8
CF-188A	X	23	Canada 2RF80-000472	188791–188798	8
CF-188B	X	24	Canada 2RF80-000472	188926–188934	9
CF-188B	X	25	Canada 2RF80-000472	188935–188940	6
			Australia		
AF-18A	VII	14	N00019-83-C-0431	A21-1–A21-3	3
AF-18B	VII	14	N00019-83-C-0431	A21-101–A21-107	7
AF-18A	VII	15	N00019-83-C-0431	A21-4–A21-7	4
AF-18A	VII	16	N00019-83-C-0431	A21-8–A21-11	4
AF-18A	VIII	17	N00019-83-C-0431	A21-12–A21-18	7
AF-18B	VIII	18	N00019-83-C-0431	A21-108–A21-112	5

Hornet Serial Numbers *(Continued)*

Type	Lot	Block	Contract Number	BuNo(s)	Quantity
			Australia		
AF-18B	VIII	19	N00019-83-C-0431	A21-113–A21-114	2
AF-18A	VIII	19	N00019-83-C-0431	A21-19–A21-21	3
			Canada		
AF-18A	IX	20	N00019-85-C-0001	A21-22–A21-27	6
AF-18A	IX	21	N00019-85-C-0001	A21-28–A21-32	5
AF-18B	IX	22	N00019-85-C-0001	A21-115–A21-116	2
AF-18A	IX	22	N00019-85-C-0001	A21-33–A21-36	4
AF-18B	X	23	N00019-85-C-0009	A21-117–A21-118	2
AF-18A	X	23	N00019-85-C-0009	A21-37–A21-40	4
AF-18A	X	24	N00019-85-C-0009	A21-41–A21-44	4
AF-18A	X	25	N00019-85-C-0009	A21-45–A21-49	5
AF-18A	XI	26	N0019-85-C-0323	A21-50–A21-53	4
AF-18A	XI	27	N00019-85-C-0323	A21-54–A21-56	3
AF-18A	XI	28	N00019-85-C-0323	A21-57	1
			Spain		
EF-18B	VIII	17	N00019-84-C-0200	CE.15-1–CE.15-2	2
EF-18A	VIII	18	N00019-84-C-0200	C.15-13	1
EF-18B	VIII	18	N00019-84-C-0200	CE.15-3–CE.15-4	2
EF-18B	VIII	19	N00019-84-C-0200	CE.15-5–CE.15-8	4
EF-18A	IX	20	N00019-85-C-0002	C.15-14–C.15-16	3
EF-18B	IX	20	N00019-85-C-0002	CE.15-9	1
EF-18A	IX	21	N00019-85-C-0002	C.15-17–C.15-21	5
EF-18B	IX	21	N00019-85-C-0002	CE.15-10–CE.15-12	3
EF-18A	IX	22	N00019-85-C-0002	C.15-22–C.15-30	9
EF-18A	X	23	N00019-85-C-0003	C.15-31–C.15-39	9
EF-18A	X	24	N00019-85-C-0003	C.15-40–C.15-45	6
EF-18A	X	25	N00019-85-C-0003	C.15-46–C.15-47	2
EF-18A	XI	26	N00019-85-C-0228	C.15-48–C.15-52	5
EF-18A	XI	27	N00019-85-C-0228	C.15-53–C.15-57	5
EF-18A	XI	28	N00019-85-C-0228	C.15-58–C.15-64	7
EF-18A	XII	29	N00019-86-C-0232	C.15-65–C.15-66	2
EF-18A	XII	30	N00019-86-C-0232	C.15-67–C.15-70	4
EF-18A	XII	31	N00019-86-C-0232	C.15-71–C.15-72	2

Hornet Serial Numbers *(Continued)*

Type	Lot	Block	Contract Number	BuNo(s)		Quantity
Ex–U.S.Navy F/A-18s Transferred to Spain*						
F/A-18A	VI	11	At North Island in September 1999	161926	C.15A-91	1
F/A-18A	VI	11	At North Island in September 1999	161935	C.15A-92	1
F/A-18A	VI	11	Transferred December 1995	161936	C.15A-73	1
F/A-18A	VI	11	Transferred March 1999	161939	C.15A-86	1
F/A-18A	VI	11	Transferred October 1996	161940	C.15A-79	1
F/A-18A	VI	11	Transferred August 1996	161944	C.15A-80	1
F/A-18A	VI	12	Transferred October 1996	161949	C.15A-81	1
F/A-18A	VI	12	Transferred May 1999	161950	C.15A-87	1
F/A-18A	VI	12	Transferred October 1997	161951	C.15A-85	1
F/A-18A	VI	12	Transferred March 1999	161953	C.15A-88	1
F/A-18A	VI	12	At North Island in September 1999	161954	C.15A-93	1
F/A-18A	VI	12	At North Island in September 1999	161958	C.15A-94	1
F/A-18A	VI	13	At North Island in September 1999	161977	C.15A-95	1
F/A-18A	VII	14	Transferred in December 1995	162415	C.15A-74	1
F/A-18A	VII	14	Transferred in December 1995	162416	C.15A-75	1
F/A-18A	VII	15	Transferred in May 1999	162421	C.15A-89	1
F/A-18A	VII	15	Transferred in December 1995	162426	C.15A-76	1
F/A-18A	VII	15	At North Island in September 1999	162444	C.15A-96	1
F/A-18A	VII	15	Transferred December 1995	162446	C.15A-77	1
F/A-18A	VII	16	Transferred August 1996	162456	C.15A-82	1
F/A-18A	VII	16	Transferred October 1996	162461	C.15A-83	1

Hornet Serial Numbers *(Continued)*

Type	Lot	Block	Contract Number	BuNo(s)		Quantity
Ex–U.S.Navy F/A-18s Transferred to Spain*						
F/A-18A	VII	16	Transferred August 1996	162465	C.15A-84	1
F/A-18A	VII	16	Transferred December 1995	162471	C.15A-78	1
F/A-18A	VII	16	Transferred May 1999	162474	C.15A-90	1
Kuwait						
KAF-18C	XIV	35	N00019-89-C-0107	401–405		5
KAF-18D	XIV	35	N00019-89-C-0107	441–443		3
KAF-18C	XIV	36	N00019-89-C-0107	406–408		3
KAF-18D	XIV	36	N00019-89-C-0107	444–446		3
KAF-18C	XIV	37	N00019-89-C-0107	409–410		2
KAF-18D	XIV	37	N00019-89-C-0107	447–448		2
KAF-18C	XV	38	N00019-89-C-0107	411–418		8
KAF-18C	XV	39	N00019-89-C-0107	419–426		8
KAF-18C	XV	40	N00019-89-C-0107	427–432		6
Switzerland						
SF-18C	XVIII	48	N00019-92-C-0057	J-5001–J-5008		8
SF-18D	XVIII	48	N00019-92-C-0057	J-5231–J-5236		6
SF-18C	XVIII	49	N00019-92-C-0057	J-5009–J-5026		18
SF-18D	XVIII	49	N00019-92-C-0057	J-5237–J-5238		2
Finland						
FN-18D	XVII	46	N00019-92-C-0140	HN-461–HN-464		4
FN-18C	XVIII	47	N00019-93-C-0051	HN-401–HN-406		6
FN-18D	XVIII	47	N00019-93-C-0051	HN-465–HN-467		3
FN-18C	XVIII	48	N00019-93-C-0051	HN-407–HN-408		2
FN-18C	XVIII	49	N00019-93-C-0051	HN-409–HN-411		3
FN-18C	XIX	50	N00019-94-C-0034	HN-412–HN-422		11
FN-18C	XX	51	N00019-95-C-0031	HN-423–HN-439		17
FN-18C	XX	52	N00019-95-C-0031	HN-440–HN-457		18
Malaysia						
MAF-18D	XIX	50	XN00019-94-C-0039	M45-01–M45-08		8

Data taken from Boeing Report ADPLS Table No. 33, F/A-18 Serial Number Conversion Chart, 25 September 1995.

*Three aircraft were scheduled to be delivered to Spain in October 1999, the last three in December 1999.

Significant Dates

1965   Air Force begins investigating an Advanced Day Fighter.

28 JANUARY 1971   Northrop unveils P530.

6 JANUARY 1972   RFP released for Light Weight Fighter (LWF) concept demonstration.

18 FEBRUARY 1972   Proposals submitted by Boeing, General Dynamics, Lockheed, Northrop, and Vought.

13 APRIL 1972   Contracts awarded to General Dynamics for YF-16 and Northrop for YF-17 demonstrators.

13 DECEMBER 1973   First YF-16 rolled out at Fort Worth.

2 FEBRUARY 1974   First YF-16 makes "official" maiden flight.

4 APRIL 1974   First YF-17 rolled at Hawthorne.

10 MAY 1974   Congress withholds Navy request for VFAX funding.

9 JUNE 1974   First YF-17 makes its maiden flight.

11 JUNE 1974   YF-17 exceeds the speed of sound in level flight without afterburner.

AUGUST 1974   Congress orders Navy to begin Navy Air Combat Fighter program.

21 AUGUST 1974   Second YF-17 makes its maiden flight.

SEPTEMBER 1974   Secretary of Defense James R. Schlesinger announced that that he was considering production of the LWF winner.

13 JANUARY 1975   YF-16 selected as winner of LWF competition.

2 MAY 1975   Navy receives permission to develop derivative of YF-17.

NOVEMBER 1975   Navy awards General Electric contract to develop F404 engine.

22 JANUARY 1976   Navy awards McDonnell Douglas contract to develop F-18, A-18, and TF-18.

13 SEPTEMBER 1978   First FSD F-18 rolled out at St. Louis.

18 NOVEMBER 1978   First FSD F-18 makes its maiden flight.

18 SEPTEMBER 1979   First FSD F-18 makes its hundredth flight.

OCTOBER 1979   Northrop sues McDonnell Douglas over F-18L.

30 OCTOBER 1979   F-18 conducts carrier qualification trials aboard USS *America.*

10 APRIL 1980   Canada announces it will purchase CF-18s.

12 APRIL 1980   First production F-18 makes its maiden flight.

13 NOVEMBER 1980   First F-18 squadron, VFA-125, is commissioned at NAS Lemoore.

20 OCTOBER 1981   Australia announces it will purchase AF-18s.

29 JULY 1982   First Canadian CF-18 makes its maiden flight.

DECEMBER 1982   Spain announces it will purchase EF-18s.

1 APRIL 1984   Department of Defense formalizes the F/A-18 designation.

6 JUNE 1984   Two St. Louis–built AF-18Bs handed over to RAAF.

26 FEBRUARY 1985   First Australian-assembled AF-18 makes its maiden flight.

APRIL 1985   Northrop versus McDonnell Douglas lawsuit settled. McDonnell Douglas becomes prime contractor for all variants of F-18 design.

4 DECEMBER 1985   First Spanish EF-18 makes its maiden flight.

APRIL 1986   F-18s conduct first combat missions during Operation Prairie Fire.

1987   Pentagon unsuccessfully tries to interest the French in a Hornet 2000 codevelopment program.

25 APRIL 1987   Blue Angels flight demonstration team conducts first public show in F/A-18s.

3 SEPTEMBER 1987   First production F/A-18C makes its maiden flight.

6 MAY 1988   First Night Attack F/A-18C makes its maiden flight.

AUGUST 1988   Kuwait announces it will purchase F/A-18C/Ds.

3 OCTOBER 1988   Switzerland announces it will purchase F/A-18C/Ds.

DECEMBER 1989   Korea announces it will purchase F/A-18C/Ds.

7 JANUARY 1991   A-12 stealth attack aircraft development program canceled.

17 JANUARY 1991   F/A-18 Hornets claim their first air-to-air victories during Operation Desert Storm.

MARCH 1991   Korea cancels F/A-18 procurement.

19 SEPTEMBER 1991   First Kuwaiti F/A-18 makes its maiden flight.

6 MAY 1992   Finland announces it will purchase F-18s.

12 MAY 1992   Navy announces F/A-18E/F Super Hornet program.

29 JUNE 1993   Malaysia announces it will purchase F/A-18Ds.

13–17 JUNE 1994   Super Hornet critical design review is held.

21 APRIL 1995   First St. Louis–assembled Finnish F-18 makes its maiden flight at St. Louis.

18 SEPTEMBER 1995   First F/A-18E is rolled out at St. Louis.

29 NOVEMBER 1995   First F/A-18E makes its maiden flight.

20 JANUARY 1996   First Swiss F/A-18 makes its maiden flight.

14 FEBRUARY 1996   First F/A-18E arrives at Pax River to begin test program.

1 APRIL 1996   First F/A-18F makes its maiden flight.

MAY 1996   Thailand announces it will purchase F/A-18C/Ds.

28 JUNE 1996   First Finnish-assembled F-18 is delivered.

JANUARY 1997   Super Hornet conducts carrier trials aboard USS *John C. Stennis.*

1 FEBRUARY 1997   First Malaysian F/A-18D makes its maiden flight.

26 MARCH 1997   F-18E/F low-rate initial production approved.

4 AUGUST 1997   The Boeing Company and McDonnell Douglas merge.

15 SEPTEMBER 1997   First production (LRIP-1) F/A-18E begins assembly in St. Louis.

8 DECEMBER 1998   Super Hornet test program surpasses the 2,000 hour mark (in F2).

13 MARCH 1998   Thailand is relieved of its commitment for F/A-18s—the U.S. will purchase the aircraft for the Marines instead.

FEBRUARY 1999   Super Hornets conduct second round of carrier trials aboard the USS *Harry S. Truman.*

MAY 1999   VX-9 begins Operational Evaluation (OPEVAL) of the Super Hornet.

NOVEMBER 1999   VX-9 completes Super Hornet OPEVAL with no major deficiencies.

Acronyms

A-X	Attack, Experimental
A/F-X	Attack/Fighter, Experimental
AAED	advanced airborne expendable decoy (ALE-50)
AAW	active aeroelastic wing
AB	Air Base
ACF	Air Combat Fighter
ACLS	automatic carrier landing system
ADF	Advanced Day Fighter
ADLP	ATARS data link pod
AEA	airborne electronic attack
AESA	advanced electronically scanned antenna
AFB	Air Force Base
AGM	air-to-ground missile (designation)
AIM	air-intercept missile (designation)
AIR	air inflatable retard (Mk 80 bomb)
AMAD	airframe-mounted accessory drive
AME	automated maintenance environment
AMRAAM	advanced medium-range air-to-air missile (AIM-120)
APU	auxiliary power unit
ARS	air-refueling system (pod)
ASD	Aeronautical Systems Division (USAF)
ASPJ	advanced self-protection jammer
ASR	advanced special receiver (ALR-67)
ASRAAM	advanced short-range air-to-air missile
ASTA	Aero-Space Technologies of Australia
ATARS	advanced tactical air reconnaissance system
ATARS-FO	advanced tactical air reconnaissance system—follow-on

ATFLIR	advanced targeting forward-looking infrared
AWACS	airborne warning and control system
BDA	battle damage assessments
BITE	built-in test equipment
BLAD	boundary layer air discharge
BuNo	Bureau Number (Navy serial number)
BVR	beyond-visual-range
C-MWS	common missile warning system
C2W	command and control warfare
CAG	Commander Air Group
CAINS	Carrier Aircraft Inertial Navigation System
CAIV	cost as an independent variable
CAP	combat air patrol
CAS	control augmentation system
CASA	Construcciones Aeronauticas SA (Spain)
CATGME	Canadian Air Task Group Middle East
CEB	Combined Effect Bomblets
CEM	combined effects munition (CBU-87)
CEP	circular error probable
CFB	Canadian Forces Base
CFD	computational fluid dynamics
CFP	concept formulation package
CIT	combined interrogator transponder (APX-111)
CRT	cathode-ray tube
CVER	canted vertical ejector rack
CW	continuous wave
DAB	Defense Acquisitions Board
DCP	development concept paper
DCS	digital communications system (Rockwell)
DEL	direct electrical link
DFRC	Dryden Flight Research Center
DoD	Department of Defense
DRI	defense reform initiatives
ECAMS	enhanced comprehensive asset management system
ECCM	electronic counter-countermeasures
ECM	electronic countermeasures
ECS	environmental conditioning system
EFA	Eurofighter 2000
EMD	engineering and manufacturing development
EPE	enhanced performance engine
ER	expanded response
ERDL	extended-range data link (Walleye/SLAM)

ERS	emergency recovery system
EW	electronic warfare
F-X	Fighter-Experimental (F-15)
F-XX	lightweight fighter, experimental
FAC	forward air control(ler)
FADEC	full-authority digital engine control
FBW	fly-by-wire
FFAR	folding-fin aircraft rocket
FIRAMS	flight incident recorder and monitoring set
FLIR	forward-looking infrared
FMS	Foreign Military Sales
FOD	foreign object damage
FOTD	fiber-optic towed decoy (ALE-55)
FPDA	Five Power Defense Agreement
FRS	fleet replacement squadron
FSD	full-scale development
FY	fiscal year (U.S.)
GAO	General Accounting Office
GCI	ground control intercept
GE	General Electric Corporation
GPS	Global Positioning System (NavStar)
HARM	high-speed antiradiation missile (AGM-88)
HARV	High Angle-of-Attack (Alpha) Research Vehicle
HATP	High Angle-of-Attack Technology Program
HOBS	high off-boresight system
HOTAS	hands-on throttle and stick
HST	HARM Targeting System
HUD	heads-up display
HUG	Hornet Upgrade (Australia)
IDECM	integrated defensive electronic countermeasures (ALQ-214)
IECMS	in-flight engine condition monitoring system
IFF	identification, friend or foe
IG	Inspector General
IIR	imaging infrared
ILS	instrument landing system
IMLC	improved multiplatform launch controller
IMP	CF-18 Incremental Modernization Project
INS	inertial navigation system
IOC	initial operational capability
IR	infrared
IRLS	infrared line scanner

JASSM	joint air-to-surface standoff missile (AGM-158)
JDAM	joint direct attack munition (GBU-29/32)
JHMCS	joint helmet-mounted cueing system
JRB	Joint Reserve Base
JSF	Joint Strike Fighter
JSIPS	joint service imagery processing station
JSOW	joint stand-off weapon (AGM-154)
LAEO	low-altitude electro-optical
lbf	pounds force
LCD	liquid crystal display
LDT/R	laser targeting designator/ranger
LDT/SCAM	laser detector tracker/strike camera (ASQ-173)
LEF	leading-edge flap
LERX	leading-edge root extension
LEX	leading-edge extension
LO	low observability
LOX	liquid oxygen
LRIP	low-rate initial production
LRU	line replaceable units
LST/SCAM	laser spot tracker/strike camera
LTD/R	laser target designator/ranger (AAS-38A/B)
LTV	Ling-Temco-Vought
LWF	Light Weight Fighter program (YF-16/YF-17)
MAEO	medium-altitude electro-optical
MALD	miniature air-launched decoy
MCAS	Marine Corps Air Station
MER	multiple ejector rack
MFD	multifunction displays
MFPG	Multinational Fighter Program Group
MIDS	multifunction information distribution system
MMH	maintenance man-hours
MPRS	multipoint refueling system
MWIR	midwave infrared
NACES	naval advanced-concept ejection seat
NACF	Navy Air Combat Fighter
NAF	Naval Air Field
NAS	Naval Air Station
NASA	National Aeronautics and Space Administration
NATC	Naval Air Test Center
NATF	Naval Advanced Tactical Fighter
NATO	North Atlantic Treaty Organization
NavAir	Naval Air Systems Command
NAVFLIR	navigation FLIR (AAR-50)

NAWCAD	Naval Air Warfare Center Aircraft Division
NFA	New Fighter Aircraft (Canada)
NITE Hawk	Navigation Infrared Targeting Equipment (AAS-38)
NORAD	North American Air Defense Command (U.S./Canada)
NSAWC	Navy Strike and Air Warfare Center
NTDS	Navy Tactical Data System
NVG	night vision goggles
NVIS	night vision intensification system
NWC	Naval Weapons Center
OBOGS	on-board oxygen generating system
OCU	Operational Conversion Unit
OPEVAL	operational evaluation
OSD	Office of the Secretary of Defense
OT&E	operational test and evaluation
PMTC	Pacific Missile Test Center
QOR	qualitative operations requirement
(RC)	reconnaissance-capable
R&D	research and development
RAAF	Royal Australian Air Force
RAF	Royal Air Force
RAIF	Research Aircraft Integration Facility
RAM	radar-absorbing material
RCAF	Royal Canadian Air Force
RCS	radar cross-section
RESCAP	rescue combat air patrol
RF	radio-frequency
RFCM	radio frequency countermeasures (ALQ-214)
RFP	request for proposals
RFQ	requests for quotations
RoKAF	Republic of Korea Air Force
RUG	radar upgrade
RWR	radar warning receiver
RWS	range-while-search (radar mode)
SACEUR	Supreme Allied Commander in Europe
SAR	synthetic aperture radar
SARH	semi-active radar homing
SASC	Swiss Aircraft and Systems Co.
SEAD	suppression of enemy aircraft defenses
SFW	sensor-fuzed weapon (CBU-97)
SHARP	shared reconnaissance pod
SRA	systems research aircraft

SRM	spin recovery mode
STT	single target track (radar mode)
SUCAP	surface combat air patrol
T-FLIR	targeting FLIR (AAS-38A/B)
TAC	Tactical Air Command
TACAN	tactical air navigation
TALD	tactical air-launched decoy (ADM-141)
TAMMC	tactical aircraft moving map capability
TAMPS	tactical mission planning system
TDP	technical development plan
TER	triple ejector rack
TFX	Tactical Fighter, Experimental (F-111)
TINS	thermal imaging navigation set (AAR-50)
TMD	tactical munitions dispenser (SUU-65)
TUDM	*Tentara Udara Diraja Malaysia* (Royal Malaysian Air Force)
TWS	track-while-scan (radar mode)
UHF	ultrahigh frequency
UN	United Nations
USAF	United States Air Force
USMC	United States Marine Corps
USN	United States Navy
VA	attack squadron (Navy)
VER	vertical ejector rack
VF	fighter squadron (Navy)
VFA	fighter-attack squadron (Navy)
VFAX	Naval Fighter-Attack, Experimental
VFMA	fighter-attack squadron (Marines)
VFX	Naval Fighter-Experimental (F-14)
VHF	very high frequency
VOR	very high frequency omnidirectional range
WVR	within-visual-range

Notes

1. This is surprising only because Air Force aircraft have traditionally not been easily adapted to carrier service, and relatively few have served in that role. Adapting naval aircraft for land-based service has proved substantially easier and more successful.

2. Dennis R. Jenkins, *McDonnell Douglas F-15 Eagle: Supreme Heavy-Weight Fighter*, Aerofax Midland Counties Publishing, 1998, pp. 6–10.

3. John W. Bohn, Jr., *Force Options for Tactical Air*, Tactical Air Command, 27 February 1965.

4. Marcelle Size Knaack, *Post-World War II Fighters: 1945–1973*, Office of Air Force History, United States Air Force, 1986, p. 334.

5. Jenkins, *McDonnell Douglas F-15 Eagle*, pp. 6–10.

6. At the time thought to be the MiG-23.

7. Counterair (air-superiority) operations are intended to achieve and maintain air superiority and, if possible, eliminate enemy air interference. Interdiction involves the reduction or elimination of support for enemy ground forces by destroying its installations and disrupting its communications. Close air support seeks to provide fire support to friendly ground forces engaged in direct combat with the enemy.

8. Jenkins, *McDonnell Douglas F-15 Eagle*, pp. 6–10.

9. Ibid.

10. Lindsay Peacock, *On Falcon Wings: The F-16 Story*, The Royal Air Force Benevolent Fund Enterprises, 1997, pp. 8–10.

11. Ibid., p. 10.

12. Jay Miller, *McDonnell Douglas F/A-18 Hornet*, Aerofax Minigraph 25, Aerofax Inc., 1988, p. 3.

13. Ibid.

14. Peacock, *On Falcon Wings*, p. 10.

15. General Dynamics would later discover that having two engines would have provided a useful redundancy, especially since the F100 would prove to have a difficult early service career, suffering many problems in both the F-15 and F-16. Nevertheless, the F-16's overall safety record is just marginally worse than the two-engine F-15 and is substantially better than any previous fighter.

16. Peacock, *On Falcon Wings*, p. 12.

17. Don Logan, *Northrop's YF-17 Cobra*, 1996, Schiffer Publishing, Ltd., p. 7.

18. This is often reported to be "unstable," but that term is too often loosely applied to vehicles with less than normal stability.

19. Miller, *McDonnell Douglas F/A-18 Hornet*, p. 2.

20. Fred Anderson, *Northrop: An Aeronautical History*, 1996, Northrop Corporation, p. 263.

21. This is how the history books record the event, but the F-15 test program later indicated that the Eagle had accomplished the feat a year earlier, but it had not been considered worthy of mention.

22. Anderson, *Northrop: An Aeronautical History*, p. 266.

23. It was not until much later that the Air Force finally allowed the F-15E to take on the interdiction mission.

24. Anderson, *Northrop: An Aeronautical History*, pp. 267–273.

25. Miller, *McDonnell Douglas F/A-18 Hornet*, p. 3.

26. Northrop Corporation, NOR-74-101, *YF-17 Flight Manual*, 10 June 1974.

27. Miller, *McDonnell Douglas F/A-18 Hornet*, p. 4.

28. Dennis R. Jenkins, *Grumman F-14 Tomcat: Leading U.S. Navy Fleet Fighter*, Aerofax Midland Counties Publishing, 1997, pp. 5–12.

29. Logan, *Northrop's YF-17 Cobra*, p. 33.

30. The P600 LWF/ACF was essentially a "stripped" P530, which had been designed and marketed as a multirole aircraft.

31. Logan, *Northrop's YF-17 Cobra*, p. 37.

32. Interestingly, the Navy did not feel compelled to redesignate the F/A-18E, which is at least as different aerodynamically from the F/A-18C as the F/A-18A was from the YF-17.

33. Robert F. Dorr, "Hornet," *World Air Power Journal*, Vol. 1, Spring 1990, p. 30.

34. The military has been confused for years over how to do this: The two-seat version of the F-102A was the TF-102A, while the two-seat F-106A was the F-106B. The two-seat F-15 was initially the TF-15A, but later became the F-15B. The F-104 was even worse: The two-seat version of the F-104A was the F-104B; the two-seat F-104G was the TF-104G.

35. The F-15 had a remarkably troublefree flight test program, and only two external changes were made to the aircraft. A "snag" was added to the stabilators to overcome a minor flutter problem, and the wingtips were reshaped to cure a minor buffeting. See Jenkins, *McDonnell Douglas F-15 Eagle*, p. 16, for a further description.

36. Dorr, "Hornet," p. 44.

37. Ford Aerospace was purchased by Loral, which was subsequently purchased by Lockheed Martin.

38. It's not bad spelling. It's an acronym that stands for Navigation Infrared Targeting Equipment.

39. Martin Marietta later merged with Lockheed to form Lockheed Martin.

40. The original F-16s used an analog fly-by-wire system. Later models have used a digital system, but the Hornet was the first fighter equipped with a fully digital system from the start.

41. Popular belief is that *fly by wire* means the control surfaces are electrically actuated. This is not the case. Current FBW aircraft retain hydraulic actuators, but command them electrically via the flight control computers instead of mechanically via direct linkage. Only recently have electric actuators become powerful, reliable, and sufficiently lightweight to be used as primary flight control actuators, although none have entered production yet.

42. Northrop Corporation, Report NB-75-301, *F-18L Tactical Fighter Weapons System Description*, June 1978.

43. The smaller fuel load did not reduce the range of the aircraft. It represented the amount of fuel necessary to carry the extra weight of the carrier gear, plus the larger reserves required by the Navy for operating at sea.

44. It is interesting to note that both snags were eventually eliminated from the F/A-18 design also.

45. Northrop Corporation, Report NB-75-301.

46. Ibid.

47. Ibid.

48. Logan, *Northrop's YF-17 Cobra*, p. 45; various *Aviation Week* and *Business Week* articles during 1980–1985.

49. Miller, *McDonnell Douglas F/A-18 Hornet*, p. 7.

50. Boeing Press Release, "Hornet: 20th Anniversary of First Flight," undated.

51. Almost universally known as "Pax River."

52. Dorr, "Hornet," p. 45.

53. F/A-18 Strike Fighter Program "Team Hornet" Newsletter, Vol. 6, No. 32, 11 August 1999.

54. Miller, *McDonnell Douglas F/A-18 Hornet*, p. 7.

55. McDonnell Douglas Press Release 80-41, 4 April 1980.

56. Miller, *McDonnell Douglas F/A-18 Hornet*, p. 9.

57. The most serious shortfall occurred at low altitudes (less than 5,000 feet) and transonic speeds (approximately Mach 0.92 to 0.97). Roll rates at slower speeds were much closer to specification. Nevertheless, the shortcoming was a major concern, since the aircraft would spend a great deal of its time at transonic speeds.

58. Miller, *McDonnell Douglas F/A-18 Hornet*, p. 8.

59. *Jane's All the World's Aircraft, 1997–98*, Jane's Information Group, Ltd., Surrey UK, p. 671.

60. Plus nine FSD aircraft, for a total of 380.

61. McDonnell Douglas Press Release 81-172, 13 October 1981.

62. It was not until the 1990s that the Navy finally began developing the "Bombcat's" ground attack capabilities, although the wing sweep control handle has always had a BOMB setting.
63. *Jane's All the World's Aircraft, 1987–88,* p. 460.
64. Miller, *McDonnell Douglas F/A-18 Hornet,* p. 11.
65. Boeing press release, "Hornet: 20th Anniversary of First Flight," undated.
66. Called the Ames-Dryden Flight Research Facility at the time—it again became DFRC in 1994.
67. Email with Dill Hunley and Peter Merlin at DFRC.
68. HARV data from the HARV fact sheets on the Dryden Web site, as well as rec.aviation.military newsgroup postings by Al Bowers, HARV's last chief engineer.
69. Email with Peter Merlin at DFRC.
70. Inconel is a nickel alloy containing chromium and iron that is extremely heat resistant and was used extensively on the X-15 and other high-speed research aircraft. Inconel is a registered trademark of Huntington Alloy Products Division of the International Nickel Company, Huntington, West Virginia.
71. Dryden Flight Research Center, *X-Press,* 17 April 1998, Internet version.
72. SRA data from Dryden fact sheets on the DFRC Web site.
73. Although Control Data won the initial Navy contract to develop and produce the first AYK-14s, many other companies have since introduced versions of the processor, all of which are interchangeable electrically and run the same software.
74. *Jane's All the World's Aircraft, 1987–88,* p. 459, and *Jane's All the World's Aircraft, 1997–98,* p. 672.
75. FY98 Annual Report, Office of the Secretary of Defense, Director of Operational Test and Evaluation.
76. U.S. Navy, manual A1-F18AC-NFM-000, Change 1, March 1985, pp. 1–5.
77. *Jane's All the World's Aircraft, 1997–98,* p. 671.
78. This includes all F/A-18s, including non-U.S. operators.
79. This points out the much more demanding environment the F/A-18 operates in compared to Air Force fighters. This loss rate is equivalent to 4.46 aircraft per 100,000 flight hours. The figure for the F-15 is 1.53, while for the F-16 it is 2.15 losses per 100,000 flight hours.
80. Bill Sweetman, "McDonnell Douglas F/A-18 Hornet," *World Air Power Journal,* Vol. 26, Autumn/Fall 1996, p. 87.
81. F/A-18 Strike Fighter Program "Team Hornet" Newsletter, Vol. 6, No. 32, 11 August 1999.
82. The Navy has a three-tiered system for classifying aircraft accidents. Class A mishaps are the most serious and involve either the loss of the aircraft or loss of life. Class B and C classifications are based on the cost to repair damages to the aircraft. VFA-125 is the first fleet replacement squadron (FRS) to complete 130,000 hours without a Class A or B mishap.
83. FY98 Annual Report, Office of the Secretary of Defense, Director of Operational Test and Evaluation.
84. Ibid.
85. *Jane's All the World's Aircraft, 1997–98,* p. 672.
86. Sweetman, "McDonnell Douglas F/A-18 Hornet."
87. *Jane's All the World's Aircraft, 1997–98,* p. 671.
88. Sweetman, "McDonnell Douglas F/A-18 Hornet," p. 77.
89. *Defense Week,* 8 November 1999, p. 16.
90. F/A-18 Strike Fighter Program *Team Hornet* Newsletter, Vol. 6, No. 32, 11 August 1999.
91. *Jane's All the World's Aircraft, 1997–98,* p. 672.
92. This was not an official designation.
93. Ibid., p. 459.
94. Ibid., p. 459.
95. Major Mark E. Merek, *Can ATARS Fix America's Tactical Reconnaissance Vacuum?,* U.S. Marine Corps Command and Staff College, 18 April 1995. "Written in Fulfillment of a Requirement for the Marine Corps Command and Staff College."
96. (RC) = reconnaissance capable. This is not an official designation, but is widely used.
97. Lieutenant Colonel David Heinz, F/A-18 Status Report, 10 December 1997.
98. *Aviation Week & Space Technology,* 24 May 1999, Internet version.
99. Ibid.
100. *Gulf War Airpower Survey,* Vol. 5; Norman Friedman, "Desert Victory," *World Air Power Journal,* p. 17.
101. General Accounting Office, Letter Report, 2 July 1996, GAO/PEMD-96-10, *Operation Desert Storm: Operation Desert Storm Air War.*
102. U.S. Navy, *War Chronology: January 1991,* U.S. Naval Historical Center, Internet version.
103. Press briefing by Lieutenant General Ray Henault, Deputy Chief of Defence Staff, 17 June 1999.
104. Lieutenant Colonel Phillip C. Tissue, USMC, "21 Minutes to Belgrade," *U.S. Naval Institute Proceedings,* September 1999, p. 38.

105. "Spanish Air Force Continued Aviano Air Ops," *Air Force News,* 20 May 1999.

106. This represents carrier-based designs only. The E-2, P-2, and P-3 land-based designs and many helicopters have done well on the export markets.

107. Mike Reyno, "Putting the Sting Back in the Hornet," *Wings: Canada's National Aviation Magazine,* Special Souvenir Edition, 1999, pp. 44–50.

108. The official Canadian designation for the aircraft is CF-188, but almost universally the aircraft is referred to as CF-18, so that term will be used here.

109. Interestingly, the name *Hornet* is not used in Canadian service because the French translation is *Frelon,* which was already used by an Aerospatiale helicopter.

110. *Jane's All the World's Aircraft, 1987–88,* p. 459.

111. "Why Does Canada Need Fighters?" *Wings: Canada's National Aviation Magazine,* Special Souvenir Edition, 1999, p. 46.

112. Andrew Cline, "Canada's Role in NORAD," *Wings: Canada's National Aviation Magazine,* Special Souvenir Edition, 1999, p. 60.

113. Reyno, "Putting the Sting Back in the Hornet," pp. 44–50.

114. In 1996, Nierlich captured the coveted "Top Gun" trophy in William Tell at Tyndall AFB, Florida—the U.S. Air Force's premier air-to-air weapons exercise. Team Canada's F/A-18s took overall first place in the competition against F-16s and F-15s.

115. Reyno, "Putting the Sting Back in the Hornet," pp. 44–50.

116. Tony Cassanova, "Operation Echo," *Wings: Canada's National Aviation Magazine,* Special Souvenir Edition, 1999, pp. 48–50.

117. These are the official serial numbers. The contractor numbers used by ASTA are prefixed with "AF" for the single-seaters and "AFT" for the two-seaters. Some sources also use AF-18 and ATF-18 as designations.

118. *Jane's All the World's Aircraft, 1987–88,* p. 459.

119. RAAF Press Release, undated, Internet version.

120. No. 3 Squadron had been based at Butterworth, Malaysia, until 31 March 1986, and received their first AF-18 two months later. No. 75 Squadron received their first Hornets in May 1988 when they were still based at RAAF Darwin. They moved to Tindal in September 1988. No. 77 Squadron received their first AF-18 on 1 July 1987.

121. *Aviation Week & Space Technology,* 4 January 1999, pp. 43–44.

122. *Australian Defence Community News,* 8 April 1999.

123. *Jane's All the World's Aircraft, 1987–88,* p. 459.

124. The "C" stands for *Caza* (Fighter) and "CE" stands for *Caza de Entrenamiento* (Fighter Trainer).

125. *Jane's All the World's Aircraft, 1997–98,* p. 673.

126. Sweetman, "McDonnell Douglas F/A-18 Hornet," p. 108.

127. *Jane's All the World's Aircraft, 1997–98,* p. 673.

128. Frank Amorosi, F/A-18 Status Report, Office of the Secretary of Defense, 1 April 1998.

129. *Jane's All the World's Aircraft, 1997–98,* p. 673.

130. U.S. Navy aircraft have traditionally been rated to $+7.5/-2.5g$ as opposed to U.S. Air Force aircraft which have been rated at $+9/-3g$ since the F-15C. Since most fourth- and fifth-generation fighters worldwide have adopted the Air Force figure as a design goal, Boeing felt it necessary to offer the capability optionally for the Hornet. The strengthening adds several hundred pounds of structural weight, and includes modified flight control software to allow the additional maneuverability. It also includes the addition of a fatigue life recording system to monitor the stresses imposed on the airframe (a technique also adopted by the F-15C when its limit was raised from the F-15A's $+7.33g$).

131. *Jane's All the World's Aircraft, 1997–98,* p. 673.

132. Amorosi, F/A-18 Status Report.

133. Later versions of the F-16 can carry FLIR pods.

134. Various press releases and background material supplied by the Finnish Air Force.

135. Finnish Air Force Headquarters, *Fighter Renewal,* 11 January 1999, pp. 4–9.

136. This shows the reliability expected from the engines, since this total includes only nine spare F404s for 64 aircraft.

137. *Jane's All the World's Aircraft, 1997–98,* p. 673.

138. *Combat Aircraft,* September 1997, p. 181.

139. *Jane's All the World's Aircraft, 1997–98,* p. 673.

140. *Aviation Week & Space Technology,* 30 March 1998, Internet version.

141. *Aviation Week & Space Technology,* 9 March 1998, p. 21.

142. *Aviation Week & Space Technology,* 23 March 1998, Internet version.

143. Ibid.

144. *Aviation Week & Space Technology,* 26 July 1999, p. 30.
145. The 1,000 aircraft figure was broken down as 660 for the Navy and 340 for the Marines. However, the Marines have never signed onto the program and it is unlikely these aircraft will be procured. This will cause the cost of the Navy aircraft to be somewhat higher.
146. General Accounting Office, *Navy Aviation: F/A-18E/F Will Provide Marginal Operational Improvement at High Cost,* Chapter Report GAO/NSIAD-96-98, 18 June 1996, Internet version.
147. This move was the subject of a separate General Accounting Office (GAO) investigation documented in *Naval Aviation: F/A-18 E/F Acquisition Strategy,* Letter Report GAO/NSIAD-94-194, 18 August 1994, Internet version.
148. *Aviation Week & Space Technology,* "Defense Officials Deny F/A-18E/F Approved Without Full Review," 6 July 1992, pp. 24–25
149. *Jane's All the World's Aircraft, 1997–98,* p. 675.
150. Barbara Starr, "McDonnell Douglas Rolls Out 'Super Hornet,' " *Jane's Defence Weekly,* 30 September 1995, p. 5.
151. General Accounting Office, *Navy Aviation: F/A-18E/F Will Provide Marginal Operational Improvement at High Cost.*
152. The aircraft could return to a carrier during the day, just barely. Navy safety rules require a 3,000-pound fuel reserve during the day; 4,000 at night.
153. Congressional Research Service, *F/A-18E/F Aircraft Program,* Paper 92035, 6 December 1996, Internet version.
154. FY98 Annual Report, Office of the Secretary of Defense, Director of Operational Test and Evaluation.
155. Federation of American Scientists Web site (http://www.fas.org).
156. U.S. Navy Public Affairs Office, "F/A-18E/F Super Hornet . . . Leading Naval Aviation into the 21st Century," 28 May 1999.
157. Navy 1999 Posture Statement.
158. *Aviation Week & Space Technology,* 3 August 1998, Internet version.
159. *Aviation Week & Space Technology,* 31 August 1998, Internet version.
160. Ibid.
161. Ibid.
162. *Aviation Week & Space Technology,* 3 August 1998, Internet version.
163. Ibid.
164. FY98 Annual Report, Office of the Secretary of Defense, Director of Operational Test and Evaluation.
165. *Aviation Week & Space Technology,* 14 June 1999, p. 77.
166. ASTA is a subsidiary of Boeing Australia, Limited.
167. Dennis R. Jenkins, *Lockheed Martin F-117 Nighthawk,* WarbirdTech Series Vol. 25, Specialty Press, 1999.
168. Sweetman, "McDonnell Douglas F/A-18 Hornet," p. 57.
169. Also called the "device"—both are equally nondescript of the classified technology.
170. Sweetman, "McDonnell Douglas F/A-18 Hornet," p. 101.
171. *Aviation Week & Space Technology,* 31 August 1998, Internet version.
172. *Jane's All the World's Aircraft, 1997–98,* p. 675.
173. Ibid.
174. Ibid.
175. FY98 Annual Report, Office of the Secretary of Defense, Director of Operational Test and Evaluation.
176. Schweizer, Roman, *Inside the Navy,* August 1999.
177. FY98 Annual Report, Office of the Secretary of Defense, Director of Operational Test and Evaluation.
178. James P. Stevenson, "Testing the Super Hornet: A Status Report," *Aerospace America,* June 1998, p. 38.
179. FY98 Annual Report, Office of the Secretary of Defense, Director of Operational Test and Evaluation.
180. *Aviation Week & Space Technology,* 31 August 1998, Internet version.
181. Patricia Frost, F/A-18 Status Report, Office of the Secretary of Defense, 1 July 1998.
182. *Aviation Week & Space Technology,* 15 March 1999, pp. 35–36.
183. Ibid.
184. FY98 Annual Report, Office of the Secretary of Defense, Director of Operational Test and Evaluation.
185. Ibid.
186. Ibid.
187. Defence Systems Internet Service, 11 August 1999.
188. *Aviation Week & Space Technology,* 30 August 1999, p. 56.
189. The Advanced Capability (AdvCap) upgrades would have significantly enhanced both the aerodynamic and electronic performance of the aging EA-6Bs. ICAP-3 is a much more modest set of upgrades, primarily aimed at the electronics.
190. *Aviation Week & Space Technology,* 26 October 1998, Internet version.

191. *Navy Times,* 16 August 1999.

192. Ever since the F-105G Wild Weasel III aircraft, the "G" designation has been informally (and sometimes officially) applied to defense suppression designs. In this case, however, it is the next logical designation after F/A-18F.

193. *Jane's All the World's Aircraft, 1997–98,* p. 675.

194. *Aviation Week & Space Technology,* 8 November 1999, p. 82; *Aviation Week & Space Technology,* 15 November 1999, pp. 81–82.

195. U.S. Navy, Manual A1-F18AA-GAI-000, September 1982.

196. This is somewhat ironic since McDonnell Douglas (Boeing) is the maker of the excellent ACES ejection seat.

197. Martin Baker calls the original seats Mk 10, while the later seats are Mk 14.

198. Miller, *McDonnell Douglas F/A-18 Hornet,* p. 15.

199. U.S. Navy, manual A1-F18AA-GAI-000, September 1982.

200. *Jane's All the World's Aircraft, 1997–98,* p. 674.

201. This rotation is referred to as *planing* (pronounced *plane-ing*). The main mounts plane when retracted and deplane during extension.

202. *Jane's All the World's Aircraft, 1997–98,* p. 674.

203. Remember, the likelihood of this occurring on a particular Hornet is once in every 3,250 hours since there are two engines. Redundancy typically increases the chances of failure, but reduces the consequence of failure.

204. Sweetman, "McDonnell Douglas F/A-18 Hornet," p. 59.

205. U.S. Navy, manual A1-F18AC-NFM-000, Change 1, March 1985.

206. Ibid.

207. Sweetman, "McDonnell Douglas F/A-18 Hornet," pp. 70–88.

208. *Jane's All the World's Aircraft, 1997–98,* p. 805.

209. Naval Air Systems Command Web site, PMA-265 F414 page, 26 July 1999.

210. *Aviation Week & Space Technology,* 17 August 1998, Internet version.

211. Ibid.

212. U.S. Navy, manual A1-F18AC-NFM-000, Change 1, March 1985, pp. 1–5.

213. *Jane's All the World's Aircraft, 1997–98,* p. 674.

214. Miller, "McDonnell Douglas F/A-18 Hornet," p. 15.

215. This is an important point since many early FBW designs did not allow this capability.

216. *Jane's All the World's Aircraft, 1997–98,* p. 676.

217. *Jane's Avionics, 1997–98,* Jane's Information Group, Ltd., Surrey UK, p. 42.

218. *Jane's All the World's Aircraft, 1997–98,* p. 674; *Jane's Avionics, 1997–98,* p. 152.

219. *Jane's Avionics, 1997–98,* p. 500.

220. *Jane's All the World's Aircraft, 1997–98,* p. 674.

221. *Commerce Business Daily,* 21 March 1996, PSA#1556

222. *Interavia,* February 1998.

223. U.S. Navy, manual A1-F18AC-742-100, Change 2, 1 September 1986.

224. *Jane's Avionics, 1997–98,* pp. 159–160.

225. The manual says to remove the LAU-116. However, many Fleet operators have discovered that only the "feet" actually need to be removed prior to mounting the sensor pods, and this is frequently done.

226. *Jane's Avionics, 1997–98,* p. 234.

227. Ibid., p. 230.

228. Ibid., p. 234.

229. U.S. Navy, Manual A1-F18AC-760-100, 1 May 1985.

230. *Jane's Avionics, 1997–98,* p. 333.

231. Ibid., p. 323.

232. At the time it was Westinghouse, which was subsequently purchased by Northrop Grumman.

233. *Jane's Avionics, 1997–98,* p. 306.

234. FY98 Annual Report, Office of the Secretary of Defense, Director of Operational Test and Evaluation.

235. Ibid.

236. *Aviation Week & Space Technology,* 26 October 1998, Internet version.

237. *Jane's Avionics, 1997–98,* p. 307.

238. Navy News Release No. 385-99, 18 August 1999.

239. *Jane's Avionics, 1997–98,* p. 307.

240. Ibid., p. 321.

241. Ibid., p. 338.

242. Sweetman, "McDonnell Douglas F/A-18 Hornet," p. 103.

243. *Aviation Week & Space Technology,* 26 October 1998, Internet version.

244. *Jane's Avionics, 1997–98,* p. 699.

245. *Jane's All the World's Aircraft, 1997–98,* p. 674.

246. Sweetman, "McDonnell Douglas F/A-18 Hornet," p. 57.

247. Although AMRAAM began as an acronym, it has essentially become the name of the missile and is often written "Amraam."

248. The Air Force had released the missile to operational use in January 1991 at the end of the Gulf War when F-15s began carrying it with the intention of shooting down SCUD missiles.

249. *Australian Defence Community News,* 4 May 1999.

250. The F-4G Wild Weasel could also use the missile in this mode, but has been retired from the inventory.

251. *Aviation Week & Space Technology,* 14 September 1998, Internet version.

252. Richard Gaskin, F/A-18 Status Report, Office of the Secretary of Defense, 1 July 1998.

253. *Aviation Week & Space Technology,* 13 September 1999, p. 84.

254. *Aviation Week & Space Technology,* 26 October 1998, Internet version.

Dennis R. Jenkins has been a senior engineer/manager in the aerospace industry for more than 20 years. For corporations such as Lockheed Martin, he has worked on projects that included the Space Shuttle, the Ballistic Missile Early Warning System, X-33/VentureStar™, and the National Airspace System (FAA). The author of several books on air- and spacecraft, including *B-1 Lancer* (another volume in the *Walter J. Boyne Military Aircraft Series*), he has degrees in both computer engineering and R&D systems management.